DIRECTORY OF

APPROVED BIOPHARMACEUTICAL PRODUCTS

DIRECTORY OF APPROVED BIOPHARMACEUTICAL PRODUCTS

Stefania Spada and Gary Walsh

CRC Press
Taylor & Francis Group
Boca Raton London New York

CRC Press is an imprint of the
Taylor & Francis Group, an **informa** business

Warning! This publication is intended for general reference only and should never be used as a guide for prescribing or administering the drugs listed.

Cover figure from:

Protein Data Bank (www.rcsb.org/pdb)
PDB ID 1HIG
Ealick, S.E., Cook, W.J., Vijay-Kumar, S., Carson, M., Nagabhushan, T.L., Trotta, P.P., Bugg, C.E.
Three-Dimensional Structure Recombinant Human Interferon-Gamma.
Science 252 pp. 698 (1991)

CRC Press
Taylor & Francis Group
6000 Broken Sound Parkway NW, Suite 300
Boca Raton, FL 33487-2742

First issued in paperback 2019

ISBN-13: 978-0-415-26368-9 (hbk)
ISBN-13: 978-0-367-39396-0 (pbk)
Library of Congress Card Number 2004050317

Library of Congress Cataloging-in-Publication Data

Spada, Stefania.
Directory of approved biopharmaceutical products / by Stefania Spada and Gary Walsh.
p. ; cm.
Includes bibliographical references.
ISBN 0-415-26368-9 (alk. paper)
1. Protein drugs—Catalogs. 2. Nucleic acids—Therapeutic use—Catalogs. I. Walsh, Gary, Dr. II. Title.
[DNLM: 1. Nucleic Acids—therapeutic use—Catalogs. 2. Pharmaceutical Preparations—Catalogs. 3. Proteins—therapeutic use—Catalogs. QV 772 S732d 2004]

RS431.P75S635 2004
615′.19--dc22 2004050317

Visit the Taylor & Francis Web site at
http://www.taylorandfrancis.com

and the CRC Press Web site at
http://www.crcpress.com

We dedicate this book to our families —

Gerard, Cristina, Maria, Nancy, Eithne, and Shane

Preface

The term "biopharmaceutical" has become an accepted part of the pharmaceutical vocabulary. It refers to therapeutic proteins produced by genetic engineering (or via hybridoma technology, in the case of some monoclonal antibodies), as well as nucleic acid-based products. By mid-2003, some 118 biopharmaceutical products had gained marketing approval and one in approximately every four genuinely new drugs coming on the market are now produced by biotechnological means.

Despite the prominence of biopharmaceuticals in the modern pharmaceutical era, relatively few publications focus, to any meaningful extent, on actual commercial products. This book aims to do just that — to provide a brief summary of each biopharmaceutical product approved for medical use. Product information is provided in monograph format. We have attempted to keep the profiles relatively consistent throughout the directory; however, some variations occur due to the information available from primary literature reference sources. The majority of information provided has been sourced from regulatory Web sites, including the European Medicines Evaluation Agency (EMEA) and the U.S. Food and Drug Administration (FDA), as well as from the Web sites of relevant pharmaceutical companies. Availability only within the E.U. and U.S. is discussed, although many of the listed products have also gained approval in other world regions. Approvals listed for E.U. products refer specifically to products approved through the new, centralized European drug approval system, introduced in 1995. Some products (particularly earlier products, such as recombinant human growth hormone preparations) gained approval in some European countries prior to 1995 and are still available in those countries.

As authors, we have ensured to the best of our ability that all the information provided is accurate. This directory, however, is intended as a general reference only and should never be used as a guide for prescribing or administering the drugs listed. Updated information (e.g., modified indications, safety warnings, etc.) on any of the drugs can be obtained by consulting the Web sites of the relevant regulatory authorities or the companies that market the products.

This directory is aimed primarily at individuals employed within the biopharmaceutical sector and at those with an interest in this sector and the products it produces. It will also serve as a useful reference for students pursuing courses in biotechnology, medicine, pharmacy, or pharmaceutical science.

Authors

Stefania Spada obtained a degree in biological sciences from the University of Milan, Italy. She engineered antibodies for carcinoma therapy at the Department of Biological and Technological Research, in the San Raffaele Scientific Institute in Milan. While working on her Ph.D. in biochemistry at the University of Zurich, Switzerland, she developed a novel phage display-based system of selection of interacting ligand pairs. Spada held a postdoctoral position at the University of Limerick, Ireland, where she identified and characterized proteins with novel biotechnological properties from thermophilic bacteria.

Gary Walsh obtained his Ph.D. from the National University of Ireland at Galway. He worked in the pharmaceutical and biotechnology industries for a number of years and is a senior lecturer of biotechnology at the University of Limerick. Walsh has published numerous articles and books on various aspects of pharmaceutical biotechnology. He sits on the editorial board of two international pharmaceutical journals and is scientific secretary of the European Association of Pharma Biotechnology.

Table of contents

Monographs are listed alphabetically by trade name

Actimmune

Product Name:	Actimmune (trade name)
	Interferon gamma-1b (international nonproprietary name)
Description:	Actimmune is a recombinant, 140-amino acid, interferon gamma-1b (IFN-γ1b). It is produced in a modified *Escherichia coli* strain and differs from the natural human molecule in containing an additional methionine residue at the N-terminal. It is supplied as a sterile solution filled in a single dose vial for subcutaneous administration. Each 0.5 ml of Actimmune contains 2 million International Units (IU) of interferon gamma-1b.
Approval Date:	1990 (U.S.). It is now also approved in various other countries.
Therapeutic Indications:	Actimmune is indicated for reducing the frequency and severity of serious infections associated with chronic granulomatous disease.
Manufacturer:	InterMune, 3294 West Bayshore Road, Palo Alto, CA 94303, http://www.intermune.com (U.S.)
Marketing:	InterMune, 3294 West Bayshore Road, Palo Alto, CA 94303, http://www. intermune.com (U.S.)

Manufacturing

Actimmune is a recombinant interferon gamma-1b. It is produced in a modified *E. coli* (strain K-12) that has been transformed with the human IFN-γ gene, and it differs from the natural human molecule by containing an additional methionine residue at its N terminus. After fermentation, the product is purified using several chromatographic and ultrafiltration steps, and the final product consists of two identical 140 amino acid chains linked via noncovalent bonds to form a homodimer. Actimmune is provided as a liquid formulation and contains mannitol, sodium succinate, and polysorbate 20 as excipients. The product is stored at 2 to 8°C. Extensive testing ensures the quality and safety of Actimmune, while its potency is assigned using a HLA-DR bioassay.

Overview of Therapeutic Properties

Actimmune is a recombinant interferon gamma-1b, which retains the biological activity of naturally occurring human interferon-γ. Interferon gamma differs from alpha and beta interferons by exhibiting a potent phagocyte-activating effect, which leads to the generation of toxic oxygen metabolites within phagocytes and results in more efficient elimination of infective microorganisms.

Actimmune is indicated for disorders characterized by deficient phagocyte oxidative activity, such as chronic granulomatous disease. Patients with this genetic condition are vulnerable to infections that may be severe and even fatal. Actimmune was found to reduce the incidence and severity of serious infections and the length of hospitalization for these patients.

More recently, Actimmune has gained approval for delaying time to progression in patients with severe, malignant osteopetrosis. This genetic disorder consists of bony overgrowth, which can lead to loss of sight and hearing, organ damage, and deficient phagocyte oxidative activity that may increase the risk of severe infections. Most patients lose sight before reaching 6 months of age and die within 10 years. Actimmune was found to delay the progression of the disease, enhance osteoclast function, and increase the 5-year survival from 30% (patients who received vitamin D, or calcitrol) to 70% (patients who received vitamin D and Actimmune).

Actimmune should be administered subcutaneously three times a week. Flu-like symptoms, headache, and reactions at the injection site were the most commonly reported side effects. Actimmune should be administered only when necessary in pregnant women and is contraindicated during lactation.

Studies involving the use of Actimmune for pulmonary disease, liver disease, and cancer are ongoing.

Further Reading

http://www.actimmune.com
http://www.fda.gov
http://www.intermune.com

Todd, P.A. and Goa, K.L., Interferon gamma-1b. A review of its pharmacology and therapeutic potential in chronic granulomatous disease, *Drugs*, 43, 111–122, 1992.

Bemiller, L.S. et al., Safety and effectiveness of long-term interferon gamma therapy in patients with chronic granulomatous disease, *Blood Cells Mol. Dis.*, 21, 239–247, 1995.

Key, L.L., Jr. et al., Long-term treatment of osteopetrosis with recombinant human interferon gamma, *N. Engl. J. Med.*, 332, 1594–1599, 1995.

Key, L.L., Jr. et al., Recombinant human interferon gamma therapy for osteopetrosis, *J. Pediat.*, 121, 119–124, 1992.

Activase

Product Name:	Activase (trade name) Alteplase (international nonproprietary name)
Description:	Activase is a first-generation recombinant human tissue plasminogen activator that promotes the degradation of fibrin and, therefore, blood clots. This 527-amino acid glycosylated serine protease is produced by recombinant DNA technology in Chinese hamster ovary (CHO) cells and is identical to the naturally occurring molecule. Activase is presented in a sterile lyophilized form in a single dose vial, containing 50 or 100 mg of the product. It is reconstituted with water for injection (WFI) before intravenous administration. Its specific activity is 580,000 IU/mg.
Approval Date:	1987 (U.S.)
Therapeutic Indications:	Activase is indicated for the management of acute myocardial infarction, acute ischemic stroke, and pulmonary embolism in adults.
Manufacturer:	Genentech, Inc., 1 DNA Way, South San Francisco, CA 94080-4990, http://www.gene.com (U.S.)
Marketing:	Genentech, Inc., 1 DNA Way, South San Francisco, CA 94080-4990, http://www.gene.com (U.S.)

Manufacturing

Activase is a recombinant human tissue plasminogen activator (tPA). The DNA encoding the product has been obtained from the natural human type I tissue plasminogen activator cDNA, derived from a human melanoma cell line. The protein is produced in CHO cells. The purification process likely includes ultrafiltration and several chromatographic steps, including anion exchange, gel filtration, and lysine affinity chromatography. The final product consists of the active substance alteplase, as well as L-arginine, polysorbate 80, phosphoric acid, and sodium chloride as excipients. The shelf life is 24 months when stored at room temperature, or 2 to 8°C.

Overview of Therapeutic Properties

Activase is a recombinant thrombolytic product. It activates the conversion of plasminogen into plasmin, which breaks down fibrin, the major component of blood clots. Activase has been largely replaced as a thrombolytic agent by the modified, long-action form, Tenecteplase (trade name). Activase is used for treatment of acute myocardial infarction to improve ventricular function, reduce congestive heart failure, and reduce mortality. Reduced mortality was observed in adult patients who received Activase within a few hours of the onset of symptoms of a heart attack. Accelerated infusion, 3-hour infusion, and two-dose regimens have been used as modes of administration.

Activase is now also indicated for the treatment of adult patients who suffer from acute ischemic stroke. Administration of Activase within 3 hours of the onset of symptoms and after cranial computer tomography (to exclude intracranial hemorrhage) led to improved neurological recovery and a reduced incidence of disability.

Activase is also indicated for the treatment of pulmonary embolism in adult patients. The most common complication related to Activase use, as with other antitrombosis agents, is the risk of bleeding — either internal or superficial — which could have severe consequences. Arrhythmias were also reported in the case of acute myocardial infarction. Activase should not be administered during pregnancy and lactation.

Further Reading

http://www.alteplase.com
http://www.fda.gov
http://www.gene.com
Albers, G.W. et al., ATLANTIS trial: results for patients treated within 3 hours of stroke onset, Alteplase Thrombolysis for Acute Noninterventional Therapy in Ischemic Stroke, *Stroke*, 33, 493–496, 2002.
Clark, W.M. et al., Recombinant tissue-type plasminogen activator (Alteplase) for ischemic stroke 3 to 5 hours after symptom onset, The ATLANTIS Study: a randomized controlled trial. Alteplase Thrombolysis for Acute Noninterventional Therapy in Ischemic Stroke, *JAMA*, 282, 2019–2026, 1999.
Ponec, D. et al., Recombinant tissue plasminogen activator (alteplase) for restoration of flow in occluded central venous access devices: a double-blind placebo-controlled trial — the Cardiovascular Thrombolytic to Open Occluded Lines (COOL) efficacy trial, *J. Vasc. Interv. Radiol.*, 12, 951–955, 2001.

Aldurazyme

Product Name:	Aldurazyme (trade name) Laronidase (common name)
Description:	Laronidase is a polymorphic variant of human α-L-iduronidase (EC 3.2.1.76), a lysosomal hydrolase that catalyses the hydrolysis of terminal α-L-iduronic acid residues of dermatan sulphate and heparan sulphate. The mature enzyme is a 628-amino acid, 83-kDa monomeric glycoprotein, containing 6 N-linked oligosaccharide side chains. It is provided as a concentrate solution for intravenous infusion. The laronidase concentration is 0.58 mg/ml, with a specific activity of approximately 172 U/mg.
Approval Date:	2003 (U.S.)
Therapeutic Indications:	Aldurazyme is indicated for patients with Hurler and Hurler-Scheie forms of mucopolysaccharidosis I (MPS I) and for patients with the Scheie form who have moderate to severe symptoms.
Manufacturer:	BioMarin Pharmaceutical Inc., 371 Bel Marin Keys Blvd., Suite 210, Novato, CA 94949, http://www.biomarinpharm.com
Marketing:	Aldurazyme is distributed in the U.S. by Genzyme Corporation, One Kendall Square, Cambridge, MA 02139, http://www.genzyme.com

Manufacturing

Laronidase is a hydrolase enzyme produced by recombinant DNA technology in an engineered Chinese hamster ovary (CHO) cell line. The producer cells are grown in suspension culture, followed by cellular harvest and recovery of the extracellular product. Multistep chromatographic purification ensues, and the production protocol includes a viral inactivation and removal step. The final product is formulated as a sterile concentrate, pH 5.5, containing sodium chloride, sodium phosphate, and polysorbate 80 as excipients. The preservative-free product is filter sterilized and aseptically filled into 5 ml single-use glass vials. Prior to infusion, the concentrate is diluted using a solution containing 0.9% sodium chloride and 0.1% human albumin in water for injection.

Overview of Therapeutic Properties

Mucopolysaccharidoses are rare, inherited genetic diseases characterized by incomplete degradation of specific mucopolysaccharides in the body. Mucopolysaccharides (glycosaminoglycans) refer to the polysaccharide side chains of a class of biological molecules known as proteoglycans. These consist of a protein backbone to which sugar side chains are attached. These polysaccharide chains can represent more than 95% of the entire molecule. When dissolved in water, proteoglycans form very viscous solutions. In the body, they usually function to support fibrous and cellular elements of tissue and help maintain water and salt balance. The major glycosaminoglycans found in the body include hyaluronic acid, chondroitin sulphate, keratan sulphate, dermatan sulphate, heparan sulphate, and heparin.

Mucopolysaccharidoses are characterized by the accumulation of the oligosaccharide component of proteoglycans due to a deficiency in one or more of the lysosomal hydrolysases normally responsible for degrading these molecules. Mucopolysaccharidosis I (MPS I) is characterized by a deficiency of α-L-iduronidase, which functions to catalyse the hydrolysis of the terminal α-L-iduronic acid residue from dermatan sulphate and heparan sulphate. As a consequence, both dermatan and heparan sulphate accumulate throughout the body, leading to widespread organ and tissue dysfunction.

Administration of Aldurazyme provides an exogenous source of the deficient enzyme. Fortuitously, much of the enzyme administered is taken up by cells and targeted to lysosomes by binding to specific mannose-6-phosphate receptors. Two of the recombinant enzyme's six oligosaccharide side chains terminate in mannose-6-phosphate residues.

The recommended dosage regimen of Aldurazyme is 0.58 mg/kg of body weight, administered once weekly by intravenous infusion. Trials have indicated that product administration improved pulmonary function and walking capacity. Liver size and urinary glycosaminoglycan levels also decreases in patients treated with Aldurazyme, although no evaluation of the product's effects on the central nervous system has been reported.

Side effects include infusion-related hypersensitivity reactions, upper-respiratory tract infections, rash, and injection-site reactions. It has not been determined if the drug is excreted in human milk. The product has been assigned a shelf life of 24 months when stored at 2 to 8°C.

Further Reading

Anon., Biomarin, genzyme begin rolling BLA for Aldurazyme: application for approval also filed in Europe, *Biotechnol. Law Rep.*, 21(4), 337–337, 2002.
Anon., Laronidase, *Biodrugs*, 16(4), 316–318, 2002.
McIntyre, J.A., Laronidase: treatment of mucopolysaccharidosis I, *Drug Future*, 28(5), 432–434, 2003.

Ambirix

Product Name:	Ambirix (trade name) Inactivated hepatitis A virus, hepatitis B surface antigen (rDNA) (common name)
Description:	Ambirix is a combined vaccine identical to the vaccine marketed under the trade name Twinrix Adult (as described in a separate monograph and administered to persons 16 years and older). It is a combination of the two separately marketed vaccines Havrix Adult, containing inactivated hepatitis A virus, and Engerix-B, containing the recombinant purified hepatitis B surface antigen. Ambirix is presented as a suspension for intramuscular administration in 1-ml prefilled syringes.
Approval Date:	2002 (E.U.)
Therapeutic Indications:	Ambirix is indicated for immunization against hepatitis A and hepatitis B in nonimmunized children and adolescents from 6 to 15 years old.
Manufacturer:	GlaxoSmithKline Biologicals, Rue de l'institut 89, 1330 Rixensart, Belgium, http://www.gsk.com (manufacturer is responsible for batch release in the European Economic Area)
Marketing:	GlaxoSmithKline Biologicals, Rue de l'institut 89, 1330 Rixensart, Belgium, http://www.gsk.com

Manufacturing

According to the Twinrix Adult monograph, the final product (per 1-ml dose) contains no less than 720 ELISA units of purified, inactivated hepatitis A virus and 20 μg of purified recombinant HBsAg protein. Excipients include aluminum, 2-phenoxyethanol, and sodium chloride. Ambirix displays a shelf life of 36 months when stored at 2 to 8°C.

Overview of Therapeutic Properties

Ambirix is identical to the product Twinrix Adult. It is a combination of Havrix Adult and Engerix-B, two previously approved vaccines against hepatitis A and hepatitis B, respectively. Ambirix differs in the volume dose from the pediatric formulation of Twinrix (Twinrix Paediatric), to which it is otherwise identical. Twinrix Paediatric is indicated for a three-dose regimen of immunization against hepatitis A and B in infants, children, and adolescents up to 15 years. Ambirix is indicated for immunization of non-immunized children and adolescents from 6 to 15 years as a two-dose regimen.

Ambirix is generally administered intramuscularly with an interval of 6 to 12 months between two injections. Ambirix was shown to elicit an immunoprotection against hepatitis A and B only 1 month after the second injection for up to 2 years. The observed response was identical to that achieved with the three-dose regimen of Twinrix Paediatric. However, it offers lower immunoprotection in the period between the two injections and should therefore be administered only in cases of low risks of infection. No studies have demonstrated a requirement for a booster administration.

The most commonly reported side effects were reactions at the site of injection, fatigue, and headache, which were slightly more severe in the case of Ambirix than Twinrix Paediatric.

Further Reading

http://www.eudra.org
http://www.gsk.com
Thoelen, S. et al., The first combined vaccine against hepatitis A and B: an overview, *Vaccine*, 17, 1657–1662, 1999.
Van der Wielen, M. et al., A two dose schedule for combined hepatitis A and hepatitis B vaccination in children ages one to eleven years, *Pediatr. Infect. Dis. J.*, 19, 848–853, 2000.

Amevive

Product Name:	Amevive (trade name) Alefacept (common name)
Description:	Alefacept is a 91.4-kDa dimeric fusion protein consisting of the extracellular CD2 binding portion of the human leukocyte function antigen 3 (LFA-3) linked to the Fc region (consisting of the hinge, C_H2 and C_H3 domains) of human IgG 1. Upon administration, the product binds to and prevents lymphocyte activation and promotes a reduction in overall lymphocyte counts. The resultant immunosuppressive action blocks the immune system cells believed to play a prominent role in the pathophysiology of chronic plaque psoriasis. The product is available in two formats; both are sterile, preservative-free, lyophilized preparations supplied in vials. Amevive for i.v. administration contains 7.5 mg active ingredient/vial, whereas Amevive for i.m. administration contains 15 mg active ingredient/vial.
Approval Date:	2003 (U.S.)
Therapeutic Indications:	Amevive is indicated for the treatment of adult patients with moderate to severe chronic plaque psoriasis who are candidates for systemic therapy or phototherapy.
Manufacturer:	Biogen Inc., 14 Cambridge Center, Cambridge, MA 02142, http://www.biogen.com
Marketing:	Biogen Inc., 14 Cambridge Center, Cambridge, MA 02142, http://www.biogen.com

Manufacturing

Amevive is produced by recombinant DNA technology in an engineered Chinese hamster ovary (CHO) cell line. After initial cell culture, the product-containing extracellular fluid is harvested. Downstream processing includes a number of high-resolution chromatographic steps. Sucrose, glycine, and citrate buffer components are added as excipients. The product is freeze dried after sterile filtration and aseptic transfer into vials.

Overview of Therapeutic Properties

Plaque psoriasis is a chronic, relapsing skin condition characterized by scaling and inflammation. Some 5.5 million people (mainly adults) are believed to suffer from psoriasis in the U.S. alone. Amevive is believed to work through an immunosuppressive action that blocks and reduces the immune system cells thought to play a major role in the disease process.

The fusion protein binds to CD2 surface antigen found primarily on T lymphocytes. Binding prevents interaction of LAF-3 (present on antigen presenting cells) and the lymphocytes. Amevive also causes a reduction in CD2+ cells, presumably by bridging between the CD2+ cell and the immunoglobulin Fc receptors found on natural killer (NK) cells and other cytotoxic cells. The cells most sensitive to Amevive are the memory effector subset of CD4+ and CD8+ T lymphocytes, which are believed to be the predominant cell types involved in psoriatic lesions.

The recommended dosage of Amevive is 7.5 mg administered once weekly as an i.v. bolus dose or 15 mg administered once weekly intramuscularly. Administration is normally over a 12-week period. CD4+ T lymphocyte counts should be monitored before and during treatment.

Clinical trials illustrated the effectiveness of Amevive in reducing the percentage of affected skin surface, as well as the severity of the scaling and inflammatory response. In addition to inducing lymphopenia, administration of Amevive may increase the risk of malignancies and the product should not be administered to patients with a history of systemic malignancy. It should be administered with caution to patients with a high risk for malignancy. Additional potentially serious side effects can include serious infections, triggered by the immunosuppressive action of the product.

Further Reading

Bos, J.D. et al., Predominance of 'memory' T cells (CD4+, CDw 29+) over naïve T cells (CD4+, CD45R+) in both normal and diseased human skin, *Arch. Dermatol. Res.*, 281, 24–30, 1989.

Craze, M. and Young, M., Integrating biological therapies into dermatological practice. Practical and economic considerations, *J. Am. Acad. Dermatol.*, 49(2), S 139–S 142, 2003.

Ellis, C. and Krueger, G.G., Treatment of chronic plaque psoriasis by selective targeting of memory effector T lymphocytes, *N. Engl. J. Med.*, 345, 248–255, 2001.
Ellis, C.N. et al., Effects of alefacept on health related quality of life in patients with psoriasis — results from a randomised placebo controlled phase II trial, *Am. J. Clin. Dermatol.*, 4(2), 131–139, 2003.
Frampton, J.E. and Wagstaff, A.J., Alefacept, *Am. J. Clin. Dermatol.*, 4(4), 277–286, 2003.
Krueger, G.G. and Callis, K.P., Development and use of alefacept to treat psoriasis, *J. Am. Acad. Dermatol.*, 49(2), 587–597, 2003.

Avonex

Product Name: Avonex (trade name)

Interferon beta-1a (international nonproprietary name)

Description: Avonex is a recombinant human interferon beta-1a (IFNβ-1a) of identical amino acid sequence to the native human molecule. The 166-amino acid, 22.5-kDa, single-chain glycoprotein is produced by recombinant DNA technology in Chinese hamster ovary (CHO) cells. It is presented as a lyophilized powder (a single-dose vial contains 30 μg interferon, equivalent to 6 million IU activity), which is resuspended using the solvent provided with a prefilled syringe. Avonex is administered intramuscularly.

Approval Date: 1996 (U.S.); 1997 (E.U.)

Therapeutic Indications: Avonex is indicated for treatment of patients with relapsing-remitting multiple sclerosis who have had at least two relapses in the preceding 3 years. Avonex is also indicated in some regions for treatment of patients at high risk of developing multiple sclerosis who have experienced a single demyelinating event.

Manufacturer: Biogen B.V., Robijnlaan 8, 2132 WX Hoofddorp, the Netherlands, http://www.biogen.com (manufacturer is responsible for import and batch release in the European Economic Area)

Biogen, Inc., 14 Cambridge Center, Cambridge, MA 02142, http://www.biogen.com (U.S.)

Marketing: Biogen France S.A., Le Capitole, 55 Avenue des Champs Pierreux, 92012 Nanterre Cedex, France (E.U.)

Biogen, Inc., 14 Cambridge Center, Cambridge, MA 02142 (U.S.)

Manufacturing

Avonex is a recombinant human interferon beta-1a. The gene, derived from human leukocytes, was amplified and cloned for production in a modified CHO cell line. The extracellular recombinant product is recovered from the culture medium and purified using several chromatographic steps and additional procedures to remove potential viral contaminants. The final product exhibits the same characteristics as the natural molecule. Avonex consists of interferon beta-1a, sodium chloride, human serum albumin, dibasic sodium phosphate, and monobasic sodium phosphate. It is provided as a lyophilized powder with a prefilled syringe, containing a solvent for reconstitution before intramuscular administration.

The shelf life of the product is 24 months when stored at a temperature of less than 25°C. Extensive tests are performed during processing and on the final product to ensure the quality and safety of the product.

Overview of Therapeutic Properties

Interferons are molecules involved in defending the body against infectious agents. They have antiviral activity, immunomodulatory activity, and inhibit cell proliferation. Interferons are effective in the treatment of multiple sclerosis, though via a poorly understood mechanism.

Avonex is indicated for the treatment of patients who have had at least two relapses in the preceding 3 years and when there is no evidence of continuous progression between the relapses. Administration of Avonex reduced the progression of the disease and decreased the exacerbation rate over a 2-year treatment period, leading to improved quality of life for patients. Avonex is also indicated in patients who have experienced a single demyelinating event with active inflammatory process and who are at high risk of developing clinically confirmed multiple sclerosis. The risk of a second event was lowered upon administration of Avonex, compared to patients who received placebos. A reduction in brain lesions was observed in most patients treated with Avonex, although it is currently unclear how brain lesions relate to disease progression.

Avonex proved to be safe when administered intramuscularly once a week, and the most commonly reported side effects were flu-like symptoms that decreased over time during the treatment. Pain at the site of injection was also reported, as well as neutralizing antibodies against interferon beta in 8% of patients who had undergone a 2-year treatment. Allergic reactions, cardiac disorders, and emotional disorders have rarely been reported. Avonex should not be administered during pregnancy and lactation, and no studies have been carried out on patients younger than 16 years old.

Further Reading

http://www.avonex.com
http://www.biogen.com
http://www.eudra.org
http://www.fda.gov

Jacobs, L.D. et al., Intramuscular interferon beta-1a therapy initiated during a first demyelinating event in multiple sclerosis, CHAMPS Study Group, *N. Engl. J. Med.*, 343, 898–904, 2000.

Vermersch, P. et al., Interferon beta1a (Avonex) treatment in multiple sclerosis: similarity of effect on progression of disability in patients with mild and moderate disability, *J Neurol.*, 249, 184–187, 2002.

BeneFIX

Product Name:	BeneFIX (trade name) Nonacog alfa (international nonproprietary name)
Description:	BeneFIX is a recombinant human coagulation factor IX involved in the blood coagulation process. The 415-amino acid, 55-kDa glycoprotein is produced in Chinese hamster ovary (CHO) cells by recombinant DNA technology and is supplied in a lyophilized form to be reconstituted for intravenous injection. Strengths varying from 250 IU to 1,000 IU are generally available.
Approval Date:	1997 (E.U. and U.S.)
Therapeutic Indications:	BeneFIX is indicated for the treatment and prevention of hemorrhagic episodes and for surgical prophylaxis in patients with hemophilia B.
Manufacturer:	Baxter S.A., Boulevard René Branquart 80, 7860 Lessines, Belgium, http://www.baxter.com (manufacturer is responsible for import and batch release in the European Economic Area) Genetics Institute, Inc., One Burtt Road, Andover, MA 01810, http://www.genetics.com (U.S.)
Marketing:	Genetics Institute of Europe B.V., Fraunhoferstrasse 15, 82152 Planegg/Martinsried, Germany (E.U.) Genetics Institute, Inc., One Burtt Road, Andover, MA 01810 (U.S.)

Manufacturing

The recombinant human coagulation factor IX (FIX) gene was cloned and expressed in a CHO cell line by recombinant DNA technology. A serum-free medium is used for the production of the recombinant FIX. The recombinant FIX is secreted in the medium and differs from the plasma-derived FIX in undergoing some post-translational modifications; these differences do not greatly affect the function of the protein.

The purification process includes four chromatographic steps and a nanofiltration step to ensure removal of viruses. The final product is presented in a lyophilized form consisting of nonacog alfa, as well as sucrose, glycine, L-histidine, and polysorbate 80 as excipients. It may also contain traces of hamster cell proteins.

The shelf life of the product is 24 months when stored at 2 to 8°C. Extensive control tests carried out during processing and on the final product to ensure quality and safety of the product include SDS-PAGE, SEC-HPLC, reverse-phase HPLC, N-terminal sequencing, MALDI-TOF MS, CD spectroscopy, and viral and bacterial analyses.

Overview of Therapeutic Properties

BeneFIX is indicated for the control and prevention of bleeding episodes in patients with hemophilia B. Hemophilia B is a hereditary disorder caused by a deficiency in coagulation factor IX, leading to inefficient blood clotting. Patients affected by hemophilia B suffer spontaneous bleeding and bleeding episodes from accidents or trauma. Human factor IX purified from plasma is commonly used in the treatment of patients with hemophilia B. Blood-derived products have the disadvantage of the potential risk of transmission of human blood-borne pathogens, which is eliminated by the use of the recombinant human factor IX produced by genetic engineering in the absence of any human-derived products.

BeneFIX is administered at individual doses specific for patient conditions to control and prevent bleeding and to control surgical bleeding. The efficacy of BeneFIX has been demonstrated in the treatment of hemophilia B patients, but higher doses may be required compared to plasma-derived human factor IX.

BeneFIX appeared to be safe. Reported side effects, though uncommon, include hypersensitivity and allergic reactions. Due to the rarity of hemophilia B (fewer than 10,000 sufferers in the E.U. and U.S. combined), clinical trials were carried out on a relatively small number of patients.

Further Reading

http://www.eudra.org
http://www.fda.gov
http://www.genetics.com

Edwards, J. and Kirby, N., Recombinant coagulation factor IX (BeneFIX), in *Biopharmaceuticals, an Industrial Perspective*, edited by Walsh, G. and Murphy, B., Eds., Dordrecht, the Netherlands: Kluwer Academic Publisher, 73–108, 1999.

Roth, D.A. et al., Human recombinant factor IX: safety and efficacy studies in hemophilia B patients previously treated with plasma-derived factor IX concentrates, *Blood*, 98, 3600–3606, 2001.

White, G.C. II et al., Recombinant factor IX, *Thromb. Haemost.*, 78, 261–265, 1997.

Beromun

Product Name: Beromun (trade name)

Tasonermin (international nonproprietary name)

Description: Beromun is a recombinant nonglycosylated cytokine of the tumor necrosis factor (TNF) alfa family (TNF α-1a). The 157-amino acid, 17.35-kDa polypeptide is produced by recombinant DNA technology in *Escherichia coli* cells. The biologically active form is a bell-shaped homotrimer. It is supplied in a lyophilized form (with single-use vials containing 1 mg of active substance) to be reconstituted and diluted before use as an intravenous infusion.

Approval Date: 1999 (E.U.)

Therapeutic Indications: Beromun is indicated for the treatment of soft tissue sarcoma of the limbs to prevent or delay amputation, or in palliative situations, for irresectable sarcoma. Beromun is used in combination with melphalan via mild hyperthermic isolated limb perfusion (ILP).

Manufacturer: Boehringer Ingelheim Austria GmbH, Dr. Boehringer-Gasse 5-11, 1121 Vienna, Austria, http://www.boehringer-ingelheim.com

Marketing: Boehringer Ingelheim International GmbH, Binger Strasse 173, 55216 Ingelheim am Rhein, Germany, http://www.boehringer-ingelheim.com

Manufacturing

The nonglycosylated cytokine tasonermin is produced by recombinant DNA technology in *E. coli*. Initial steps in the purification process include lysis and homogenization of the cell culture, followed by precipitation with polyethyleneimine to remove DNA and some contaminant proteins. This is followed by a series of five chromatographic steps, apparently involving ion exchange, hydrophobic interaction, and gel filtration chromatography. The final purified product is presented in a lyophilized form consisting of tasonermin, as well as sodium dihydrogen phosphate dihydrate, disodium hydrogen phosphate dodecahydrate, sodium chloride, and human serum albumin as excipients.

The shelf life of the product is 36 months when stored at 2 to 8°C. Extensive control tests are carried out to ensure quality and safety of the product, including DNA agarose gel analysis, restriction mapping, DNA sequence analysis, SDS-PAGE, and ELISA analysis.

Overview of Therapeutic Properties

Soft tissue sarcoma is a tumor with low incidence. In 60% of patients it is found in the limbs. Beromun is indicated for the treatment of 10% of those patients who generally undergo amputation or severe surgery.

The exact mechanism of action of Beromun is not known, but the cytokine is a toxic target for the vasculature of the tumor and induces tissue necrosis. It also has a modulating effect on the immune response.

Beromun is administered only in specialized locations via ILP, a technique that allows local administration of the drug, avoiding circulation through the whole body. Radioactive labeling is used to monitor leakage to the body. A leakage of higher than 10% will induce severe side effects. Beromun is administered as an infusion (at a dose of 3 mg for treatment of the arm and 4 mg for the leg) over 30 minutes under mild hyperthermia, which increases the efficacy of the treatment. Administration of melphalan follows, with the whole treatment lasting 90 minutes. If a second treatment is required, it cannot be performed within 6 weeks of a previous treatment.

Clinical trials established the efficacy of Beromun, with 60% of patients experiencing a better outcome than predicted over an average period of 500 days. No effect was observed on overall survival.

The most common side effects reported were fever, nausea, vomiting, chills, heart and pressure disorders, kidney and liver toxicity, fatigue, and infections. Common local reactions in the affected limb were pain, swelling, and damage to the skin, nerves, and nails. In 2% of patients, the severe tissue damage led to amputation. Serious side effects reported were liver toxicity and thrombocytopenia. Beromun is contraindicated in the case of heart or circulatory disorders, lung, liver, or severe kidney diseases, infections not responding to antibiotics, severe fluid buildup, pregnancy, and lactation. It should not be administered if patients cannot make use of vasopressors, anticoagulants, or radioactive tracers. Safety and efficacy has not been studied for patients younger than 16 years old.

Further Reading

http://www.boehringer-ingelheim.com
http://www.eudra.org
Hoekstra, H.J. et al., Continuous leakage monitoring during hyperthermic isolated regional perfusion of the lower limb: techniques and results. *Reg. Cancer Treat.*, 4, 301–304, 1992.

Betaferon/Betaseron

Product Name:	Betaferon (trade name in E.U.) and Betaseron (trade name in U.S.) Interferon beta-1b (international nonproprietary name)
Description:	Betaferon (Betaseron) is a recombinant human interferon beta that differs from the native protein in that it has a serine at position 17 (replacing a cysteine), no methionine at position 1, and is not glycosylated. The 18.5-kDa protein is produced by recombinant DNA technology in *Escherichia coli* and supplied in a lyophilized form to be reconstituted to a strength of 0.25 mg/ml before administration as a subcutaneous injection.
Approval Date:	1993 (U.S.); 1995 (E.U.)
Therapeutic Indications:	Betaferon is indicated under certain conditions for relapsing-remitting and secondary progressive multiple sclerosis
Manufacturer:	Schering AG, D-13342 Berlin, Germany, http://www.schering.de (manufacturer is responsible for import and batch release in the European Economic Area) Chiron Corporation, 4560 Horton Street, Emeryville, CA. http://www.chiron.com (U.S.)
Marketing:	Schering AG, D-13342 Berlin, Germany, http://www.schering.de (E.U.) Berlex Inc., 340 Changebridge Road, P.O. Box 1000, Montville, NJ, 07045-1000, http://www.berlex.com (U.S.)

Manufacturing

The interferon beta-1b was generated by recombinant DNA technology from the native human interferon beta gene. The genetic sequence of the recombinant product was modified by site directed mutagenesis in order to replace a cysteine at position 17 in the native protein with a serine. The elimination of the -SH group has the advantage of avoiding incorrect disulfide bridge formation within the molecule and intermolecular crosslinking during the manufacturing process. The methionine residue at position 1 is removed in the recombinant molecule, and, expressed in a prokaryotic system, it is not glycosylated. Nevertheless, functionality and immunogenicity of the protein are not significantly affected. The recombinant protein is produced in *E. coli* cells, and during production soya bean extract is used instead of bovine-derived products. Interferon beta-1b is extracted from inclusion bodies following homogenization and centrifugation of the harvested cells. Purification involves multiple chromatographic steps (mainly multiple gel filtration steps) and filtration steps. The final product is presented in a lyophilized form consisting of interferon beta-1b, as well as human albumin and dextrose as excipients. A sodium chloride solution is provided for reconstituting the product before administration.

The shelf life of the product is 18 months when stored at 2 to 8°C. Routine evaluative tests are carried out on the final product to ensure its quality and safety.

Overview of Therapeutic Properties

Multiple sclerosis (MS) is a chronic, disabling disease of the central nervous system. In the benign form (15% of cases) no disabilities occur, whereas primary progressive MS (10 to 15% of cases) brings minor loss of function. The majority of patients (70%) suffer from relapsing-remitting and secondary progressive disease and develop significant disabilities — either gradually or due to the relapsing-remitting, which involves a worsening of the disease followed by partial or complete recovery.

Betaferon is indicated for patients with relapsing-remitting disease who have suffered two or more relapses in the preceding 2 years and for patients with secondary progressive MS.

The mechanism of action of Betaferon is not completely understood, but it is mediated by specific cell receptors and involves the modulation of the immune response.

Betaferon is administered as a subcutaneous injection every second day at a recommended dose of 0.25 mg over a 2- or 3-year period. Betaferon treatment has been shown to reduce the frequency and severity of relapses, increasing the interval between relapses and reducing the progression of the disease. Hospitalization of MS patients decreased after Betaferon treatment, compared to patients receiving placebos, and a delay was observed in patients becoming wheelchair bound.

The most common side effects of Betaferon are the occurrence of injection-site necrosis (so it is important to rotate the injection site regularly) and flu-like symptoms. Rare but severe hypersensitivity reactions have been observed, as well as muscular hypertonia and rare cases of cardiomyopathy. After 18 to 24 months of treatment, a reduction of efficacy may be observed due to the development of neutralizing antibodies against Betaferon. Suicidal thoughts and depressive disorders have been reported, but they are not necessarily related to the administration of interferon beta-1b. Betaferon is contraindicated during pregnancy and lactation. It should not be administered to patients younger than 18 years old.

Further Reading

http://www.betaseron.com
http://www.chiron.com
http://www.eudra.org
http://www.fda.gov
http://www.schering.de
Dhib-Jalbut, S. and McFarland, H.F., Treatment of multiple sclerosis with interferon beta 1b, *Baillieres Clin. Neurol.*, 6, 467–480, 1997.
Kelley, C.L. and Smeltzer, S.C., Betaseron: the new MS treatment, *J. Neurosci. Nurs.*, 26, 52–56, 1994.

Bexxar

Product Name: Bexxar (trade name)

Tositumomab (common name)

Description: Bexxar contains tositumomab, a murine IgG_{2a} monoclonal antibody that specifically binds the CD 20 antigen. CD 20 is a surface antigen found on B lymphocytes, both normal and malignant. The 150-kDa antibody is composed of two gamma 2a heavy chains (451 amino acids each) and two light (lambda) chains (220 amino acids each). The Bexxar therapeutic regime entails the use of both tositumomab and radioactive tositumomab. Tositumomab is supplied as a sterile, preservative-free liquid concentrate containing 14 mg active/ml in 35-mg and 225-mg single-use vials. Radiolabeled tositumomab (^{131}I- tositumomab) is also supplied as a sterile, preservative-free liquid concentrate, available at concentrations of 0.1 and 1.1 mg/ml.

Approval Date: 2003 (U.S.)

Therapeutic Indications: Bexxar is indicated for the treatment of patients with CD 20 positive, follicular, non-Hodgkin's lymphoma (NHL), with and without transformation, whose disease is refractory to Rituximab and has relapsed following chemotherapy. It is not indicated for the initial treatment of patients with CD 20 positive NHL.

Manufacturer: Corixa Corporation, Seattle, WA 98104, http://www.corixa.com

Marketing: Co-marketed in the U.S. by Corixa Corporation, Seattle, WA 98104 and GlaxoSmithKline, Philadelphia, PA 19101

Manufacturing

Tositumomab is manufactured by culture of the antibody-producing mammalian cell line in antibiotic-free cell media. After cell removal, the antibody-containing extracellular fluid is subjected to a number of downstream processing steps, including filtration and high-resolution chromatographic purification. The profile and exact concentration of excipients added varies. Non-radiolabeled tositumomab preparations contain maltose and sodium chloride, as well as phosphate buffer components, and display a pH of approximately 7.2. The radiolabeled product contains povidone, maltose, sodium chloride, and ascorbic acid as excipients. Its final pH is approximately 7.0.

Radiolabeled tositumomab is manufactured by direct chemical coupling of ^{131}I to the antibody. Unbound iodine and other reactants are then removed from the radiolabeled antibody by a gel filtration step.

Vials of tositumomab should be stored at 2 to 8°C, protected from light. Vials of the radiolabeled tositumomab should be stored frozen (–20°C or below) in the protective lead pots in which they were supplied.

Overview of Therapeutic Properties

Tositumomab binds specifically to human B lymphocyte-restricted differentiation antigen, better known as CD 20 antigen. CD 20 is a transmembrane phosphorylated protein expressed on pre-B lymphocytes and, at higher density, on mature B lymphocytes. It is also expressed on more than 90% of B cell NHLs. NHL is currently the sixth leading cause of cancer-related deaths in the U.S., killing almost 25,000 Americans each year. Administration of Bexxar results in a sustained depletion of circulating CD 20 positive cells, both transformed and nontransformed. Binding of the antibody to the extracellular portion of the CD 20 may trigger cell death by a number of mechanisms, including induction of apoptosis, antibody-dependent cellular cytotoxicity, complement-mediated effects, as well as ionizing radiation derived from ^{131}I-labeled antibody. The radioactive iodine (half-life of 8 days) is ultimately eliminated through the urine.

The recommended Bexxar therapeutic regime entails administration of both unlabeled and ^{131}I-labeled forms of the antibody to patients and is intended as a single-course treatment. The regime consists of four component injections administered in two steps — the dosimetric step and the therapeutic step. The dosimetric step entails initial i.v. infusion of 450 mg tositumomab (unlabeled) in a volume of 50 ml over 60 minutes. This step is followed by i.v. infusion of 35 mg of radiolabeled (50 mCi ^{131}I) tositumomab over 20 minutes. This is followed by whole body dosimetry and biodistribution studies using whole body images acquired with a gamma camera. If biodistribution is acceptable, therapeutic doses can be administered, usually 6 to 7 days after the dosimetric dose. Therapeutic dosage entails initial i.v. administration of 450 mg tositumomab (unlabeled in 50 ml final volume over 60 minutes), followed by i.v. administration of radiolabeled tositumomab, the exact quantity of which is determined by the results of the dosimetry steps.

Administration of Bexxar can induce a number of serious, sometimes fatal, side effects. Hypersensitivity reactions, including anaphylaxis, may occur upon product infusion. Most patients receiving the therapeutic regime develop severe thrombocytopenia and neutropenia. The product should not be administered to pregnant women as the ^{131}I may harm the fetal thyroid gland. Among other serious adverse effects was the development of secondary malignancies (myelodysplastic syndrome and acute leukemia) in a number of patients.

Further Reading

Anon., Bexxar — tositumomab and I 131 tositumomab. GlaxoSmithKline/Corixa — new treatment for non-Hodgkin's lymphoma, *Formulary*, 38(8), 455–456, 2003.

Anon., Iodine-131 tositumomab, *Biodrugs*, 17(4), 290–295, 2003.

Berdeja, J.G., Immunotherapy of lymphoma. Update and review of the literature, *Curr. Opin. Oncol.*, 15(5), 363–370, 2003.

Blum, K. and Bartlett, N., Antibodies for the treatment of diffue large cell lymphoma, *Semin. Oncol.*, 30(4), 448–456, 2003.

Cardarelli, P.M. et al., Binding to CD 20 by anti-B1 antibody or F (ab) (2) is sufficient for induction of apoptosis in B cell lines, *Cancer Immunol. Immunother.*, 51(1), 15–24, 2002.

Press, O.W. et al., Retention of B-cell specific monoclonal antibodies by human lymphoma cells, *Blood*, 83, 1390–1397, 1994.

Bio-Tropin (Tev-Tropin or Zomacton)

Product Name: Bio-tropin, Tev-tropin, Zomacton (trade names)

Somatropin (international nonproprietary name)

Description: Bio-tropin (Tev-tropin in U.S. or Zomacton in E.U.) is a recombinant human growth hormone (hGH, somatropin). The 22.1-kDa, 191-amino acid, single-chain polypeptide is produced in a modified *Escherichia coli* strain using recombinant DNA technology.

Approval Date: Bio-tropin was first approved in 1995 in the U.S., although it has not been marketed thus far in the U.S. due to a patent dispute. The product was developed by Biotechnology General Corporation, now known as Savient Pharmaceuticals. Savient intends to launch the product in the U.S., where it will be distributed by Teva Pharmaceuticals, under the trade name Tev-tropin. The product has been marketed for a number of years outside the U.S. in several world regions. It is marketed in Europe by Ferring Pharmaceuticals under the trade name Zomacton.

Therapeutic Indications: Zomacton is indicated in Europe for the treatment of children with a growth failure due to insufficient endogenous growth hormone or due to Turner's syndrome.

Manufacturer: Teva Pharmaceuticals, 5 Basel Street, P.O. Box 3190, Petah Tikva 49131, Israel

Marketing: Teva Pharmaceuticals, 1090 Horsham Road, P.O. Box 1090, North Wales, PA 19454 (U.S.)
Savient Pharmaceuticals Inc., 1 Tower Center, 14th Floor, East Brunswick, NJ 08830, http://www.savientpharma.com (U.S.)
Ferring Pharmaceuticals, various European countries, http://www.ferring.com (E.U.)

Manufacturing

This product is produced by recombinant DNA technology in *E. coli* cells. After the initial fermentation step, the hormone is purified using a number of chromatographic steps. The final filter-sterilized product is filled into vials and presented in lyophilized form (with 4 mg active ingredient and mannitol as an excipient) with a solvent for product reconstitution. The solvent consists of water for injection containing sodium chloride and benzyl alcohol.

Overview of Therapeutic Properties

Clinical data support the efficacy of Zomacton in the treatment of children with a growth failure due to insufficient endogenous growth hormone or due to Turner's syndrome.

In the case of growth hormone deficiency, a dose of 0.17 to 0.23 mg/kg body weight per week (administered through 6 or 7 SC injections) is recommended. In the case of Turner's syndrome, a dose of 0.33 mg/kg body weight per week (administered through 6 or 7 SC injections) is recommended. The product is contraindicated in patients with closed epiphyses; pregnant and lactating mothers; patients with evidence of active neoplasm; patients with acute critical illness suffering complications following multiple accidental trauma, open heart, or abdominal surgery, or acute respiratory failure; and patients with a known hypersensitivity to any product components. Other potential side effects include insulin resistance and hypothyroidism.

Further Reading

Agerso, H. et al., Pharmacokinetics and pharmacodynamics of a new formulation of recombinant human growth hormone. Administration by ZomaJet 2 Vision, a new needle-free device compared to subcutaneous administration using a conventional syringe, *J. Clin. Pharmacol.*, 42(11), 1262–1268, 2002.

Anon., Genentech blocks U.S. sales of BTGS Biotropin, *Eur. Chem. News*, 64(1681), 19, 1995.

Zadik, Z. et al., Effect of timing of growth hormone administration on plasma growth hormone binding activity, insulin like growth factor-1 and growth in children with a subonormal spontaneous secretion of growth hormone, *Horm. Res.*, 39(5-6), 188–191, 1993.

Bioclate

Product Name:	Bioclate (trade name) Antihemophilic factor (recombinant) (international nonproprietary name)
Description:	Bioclate consists of a recombinant human coagulation factor VIII, the recombinant antihemophilic factor (rAHF). It appears to be identical to the product Recombinate, marketed by Baxter. It is produced in Chinese hamster ovary (CHO) cells using recombinant DNA technology. It is provided as a lyophilized powder (containing 250, 500, or 1000 IU factor VIII activity per vial) to be reconstituted before intravenous administration.
Approval Date:	1993 (U.S.)
Therapeutic Indications:	Bioclate is indicated for the prevention and control of bleeding in patients suffering from hemophilia A. It is not indicated in von Willebrand's disease.
Manufacturer:	Baxter Healthcare Corporation, Hyland Immuno, Glendale, CA 91203, http://www.baxter.com
Marketing:	Aventis Behring, 1020 First Avenue, King of Prussia, PA 19406, http://www.aventisbehring.com

Manufacturing

The rAHF is expressed (as an extracellular glycoprotein) in an engineered CHO cell line. It is purified from the cell culture medium by a series of chromatographic steps, including ion exchange, gel filtration, and a high-resolution immunoaffinity step employing an antifactor VIII monoclonal antibody. It is formulated as a lyophilized powder containing human albumin, polyethylene glycol, histidine, polysorbate 80, and calcium as excipients.

Extensive tests are carried out to validate the potency, identity, purity, sterility, viral safety, and specific activity of the final product.

Overview of Therapeutic Properties

Patients suffering from hemophilia A lack a functional factor VIII and require replacement therapy in order to avoid and control bleedings. Bioclate is indicated for hemophilia A and is generally administered intravenously at a dosage adjusted to individual needs. The product was shown to be as effective as the natural human coagulation factor VIII in the management of hemophilia A in patients who had previously been treated with plasma factor VIII, in untreated patients, and in the prophylaxis of surgical procedures.

It was found to be safe with a very low incidence of side effects, mostly infusion-related. Allergic reactions, due to traces of murine, bovine, and hamster proteins, were rarely reported. Neutralizing antibodies that interfere with the efficacy of the product have been reported.

Further Reading

http://www.baxter.com

http://www.fda.gov

Belgauini, A. et al., Stability and sterility of a recombinant factor VIII concentrate (Bioclate™) prepared for continuous infusion prophylaxis, *Thromb. Haemostasis.*, 2092–2092, 1997.

Belgaumi, A. et al., Stability and sterility of a recombinant factor VIII concentrate prepared for continuous infusion administration, *Am. J. Hematol.*, 62(1), 13–18, 1999.

Deitcher, S. et al., Intranasal DDAVP induced increases in plasma von Willebrand factor alter the pharmacokinetics of high-purity factor VIII concentrates in severe haemophilia A patients, *Haemophilia*, 5(2), 88–95, 1999.

CEA-Scan

Product Name: CEA-Scan (trade name)

Arcitumomab (international nonproprietary name)

Description: CEA-Scan consists of the (antigen-binding) Fab' fragment of a monoclonal antibody raised against the human carcinoembryonic antigen (CEA), which is mainly found in association with various carcinomas such as colorectal carcinomas. CEA-Scan is produced in mice by ascites technology and is provided in a lyophilized form (1.25 mg/vial). CEA-Scan is conjugated with radioactive technetium (^{99m}Tc) immediately before administration as an intravenous injection.

Approval Date: 1996 (E.U. and U.S.)

Therapeutic Indications: CEA-Scan is indicated for imaging recurrence and metastases in patients with histologically demonstrated colorectal carcinoma as an adjunct to standard noninvasive techniques.

Manufacturer: Eli Lilly Pharma Fertigung und Distribution GmbH & Co. KG, Teichweg 3, 35396 Giessen, Germany, http://www.lilly.com (manufacturer is responsible for import and batch release in the European Economic Area)
Immunomedics, Inc., 300 American Road, Morris Plains, NJ 07950, http://www.immunomedics.com (U.S.)

Marketing: Immunomedics Europe, Haarlemmerstraat 30, 2181 HC Hillegom, Netherlands, http://www.immunomedics.com (E.U.)
Immunomedics, Inc., 300 American Road, Morris Plains, NJ 07950, http://www.immunomedics.com (U.S.)

Manufacturing

Lymphocytes from mice immunized with human CEA were fused with mouse myeloma cells to generate the hybridoma cell line producing the desired monoclonal antibody (known as Immu-4). The hybridoma cells are injected into pristane-primed mice and ascites, containing the monoclonal anti-CEA antibody, is removed 14 to 30 days after injection. The antibody (purified using a combination of two ion exchange and one affinity chromatographic step) is digested with pepsin to produce $F(ab')_2$ antibody fragments, followed by cysteine reduction to Fab'-SH. Additional purification steps are then undertaken to remove undigested antibody, additional fragments and pepsin. The final product is presented in a lyophilized form consisting mainly of the anti-CEA Fab'-SH but also containing $F(ab')_2$ and H- and L-chain fragments and the following excipients: stannous chloride, sodium chloride, sucrose, argon, sodium potassium tartrate, sodium acetate, and traces of acetic acid and hydrochloric acid for a final buffered solution at pH 5 to 7.

Evaluative tests were carried out to verify the purity of the final product and the quality of the intermediates (i.e., HPLC, IEF, GC, SDS-PAGE, ITLC, and DNA hybridization), the potency of the product (immunoreactivity assay), microbiological and viral safety (mycoplasma assay, XC plaque assay, S^+L^- focus-forming assay, reverse transcriptase assay, and electron microscopy), and the presence of endotoxins and pyrogens. The shelf life of the product is 48 months when stored at 2 to 8°C.

Overview of Therapeutic Properties

Following reconstitution with the radioisotope technetium-99m, CEA-Scan can be administered to patients with histologically proven colon or rectum carcinoma for radioimaging to detect tumor recurrence or metastases. The Fab' antibody fragment recognizes the CEA, which is a tumor marker for carcinomas. CEA-Scan is used in combination with other diagnostic methods, such as computerized tomography (CT).

CEA-Scan is administered as an intravenous injection followed by radioimaging within 2 to 5 hours using a standard nuclear camera. The use of CEA-Scan in combination with CT, and in patients with proven colorectal carcinoma, overcomes the fact that arcitumomab may cross react with other cell types.

The use of the Fab' fragment arcitumomab has the following advantages over the use of a whole IgG molecule: its rapid clearance from the blood, with a half-life of approximately 6 hours, enables imaging after a shorter tumor localization period, and thus exposes the patient to less radiation; the

Fab' molecule is less immunotoxic and immunogenic than a whole IgG, with fewer than 1% of patients developing a human antimouse antibody (HAMA) response in clinical trials, making readministration possible; and the Fab' molecule, with free thiol groups, is easily conjugated to the radioisotope.

CEA-Scan proved to be safe and well tolerated, with side effects being uncommon, mild, and mostly transient. CEA-Scan is contraindicated during pregnancy and lactation. No safety data are available for patients under 21 years of age. Ongoing studies are monitoring the use of CEA-Scan in diagnoses of other tumors, such as breast and lung tumors.

Further Reading

http://www.cea-scan.com
http://www.eudra.org
http://www.fda.gov
http://www.immunomedics.com

Moffat, F.L., Jr. et al., Clinical utility of external immunoscintigraphy with the IMMU-4 technetium-99m Fab' antibody fragment in patients undergoing surgery for carcinoma of the colon and rectum: results of a pivotal, phase III trial. The Immunomedics Study Group, *J. Clin. Oncol.*, 14, 2295–2305, 1996.

Wegener, W.A. et al., Safety and efficacy of arcitumomab imaging in colorectal cancer after repeated administration, *J. Nucl. Med.*, 41, 1016–1020, 2000.

Cerezyme

Product Name: Cerezyme (trade name)

Imiglucerase (international nonproprietary name)

Description: Cerezyme is a recombinant modified (macrophage targeted) human β-glucocerebrosidase enzyme. It is produced by recombinant DNA technology in Chinese hamster ovary (CHO) cells and is supplied in a lyophilized form (200 IU activity per vial) to be reconstituted and diluted before administration by intravenous infusion.

Approval Date: 1994 (U.S.); 1997 (E.U.)

Therapeutic Indications: Cerezyme is indicated for use as long-term enzyme replacement therapy in patients with certain forms of Gaucher's disease.

Manufacturer: Genzyme Ltd., 37 Hollands Road, Haverhill, Suffolk CB9 8PU, UK, http://www.genzyme.com (manufacturer is responsible for import and batch release in the European Economic Area)
Genzyme Corporation, One Kendall Square, Cambridge, MA 02139-1562, http://www.genzyme.com (U.S.)

Marketing: Genzyme B.V., Gooimeer 10, 1411 DD Naarden, the Netherlands, http://www.genzyme.com (E.U.)
Genzyme Corporation, One Kendall Square, Cambridge, MA 02139-1562, http://www.genzyme.com (U.S.)

Manufacturing

The recombinant human β-glucocerebrosidase differs from the native form by one amino acid, a substitution of arginine for histidine at position 495. It also demonstrates an altered glycosylation pattern. The protein is produced in a CHO cell line. The manufacturing process includes an *in vitro*, post-translational enzymatic processing step using an exoglycosidase. This removes the sialic acid sugars that cap the product's oligosaccharide side chains, exposing the mannose residues underneath. Exposure of mannose in turn facilitates selective uptake of the product by macrophages, due to the presence of mannose receptors on the surface of the latter. The product is purified by a combination of chromatographic steps and is presented in a lyophilized form containing mannitol, sodium citrate, citric acid monohydrate, and polysorbate 80 as excipients.

The shelf life of the product is 24 months when stored at 2 to 8°C. Control tests are carried out during processing and on the final product to ensure its quality and safety.

Overview of Therapeutic Properties

Gaucher's disease is a genetic disease resulting from a deficiency of the lysosomal enzyme glucocerebrosidase, which breaks down the glycolipid glucocerebroside. The disease is characterized by accumulation of glucocerebroside in tissue macrophages, which become engorged and are generally termed Gaucher's cells. Gaucher's cells accumulate in spleen, liver, and bone marrow, resulting in anemia and thrombocytopenia. Administration of the deficient enzyme is used to treat Gaucher's disease. The recombinant enzyme imiglucerase differs from the native form in its glycosylation pattern. The exposure of mannose residues enables the selected uptake of the enzyme by macrophages. Cerezyme replaces the human placental-derived enzyme, alglucerase, with the advantage of reduced risk of viral contamination. Cerezyme is administrated as an intravenous infusion, usually every 2 weeks.

Clinical trials proved its efficacy in the treatment of Gaucher's disease, exhibiting improved hematological values and reduction of spleen and liver volume. Cerezyme showed no difference in efficacy compared to the placental-derived product. Due to the rarity of the disease, clinical trials were performed on a small number of patients.

Clinical trials proved to be safe and well tolerated. In 15% of patients, antibodies against Cerezyme developed, but the efficacy of the product was not affected. Rare severe hypersensitivity reactions, pulmonary hypertension, anaphylactoid reactions, tachycardia, and cyanosis have been reported. Cerezyme is contraindicated during pregnancy and lactation.

Further Reading

http://www.cerezyme.com
http://www.eudra.org

http://www.fda.gov
http://www.genzyme.com
Grabowski, G.A. et al., Enzyme therapy in type 1 Gaucher's disease: comparative efficacy of mannose-terminated glucocerebrosidase from natural and recombinant sources, *Ann. Intern. Med.*, 122, 33–39, 1995.
Hoppe, H., Cerezyme — recombinant protein treatment for Gaucher's disease, *J. Biotechnol.*, 76, 259–261, 2000.
Niederau, C., First long-term results of imiglucerase therapy of type 1 Gaucher's disease, *Eur. J. Med. Res.*, 3, 25–30, 1998.

Comvax

Product Name:	Comvax (trade name) Hemophilus b conjugate (meningococcal protein conjugate) and hepatitis B (recombinant) vaccine (common name)
Description:	Comvax is a bivalent vaccine that contains antigenic components derived from two already marketed vaccines: Pedvax HIB (hemophilus b conjugate vaccine) and Recombivax HB (recombinant hepatitis B vaccine). The components are the *Haemophilus influenzae* type b (Hib) capsular polysaccharide (PRP) covalently bound to an outer membrane protein complex of *Neisseria meningitidis* (OMPC) and recombinant hepatitis B surface antigen (rHBsAg). The product is generally supplied in a box containing 10 single-dose (0.5 ml) vaccine vials and is administered intramuscularly.
Approval Date:	1996 (U.S.)
Therapeutic Indications:	Comvax is indicated for vaccination against invasive disease caused by Hib and against infection caused by all known subtypes of hepatitis B virus in infants 6 weeks to 15 months of age born of HBsAg-negative mothers.
Manufacturer:	Merck and Co., Inc., One Merck Drive, Whitehouse Station, NJ 08889-1000, http://www.merck.com
Marketing:	Merck and Co., Inc., One Merck Drive, Whitehouse Station, NJ 08889-1000, http://www.merck.com

Manufacturing

Hib, *Neisseria meningitidis,* and recombinant *Saccharomyces cerevisiae* are all produced separately by fermentation. The Hib capsular polysaccharide, PRP, is purified from relevant culture broth by a series of steps including ethanol fractionation, enzymatic digestion, phenol extraction, and diafiltration. The outer membrane protein complex, OMPC, from *N. meningitidis* is purified by a combination of detergent extraction, ultracentrifugation, diafiltration, and sterile filtration. The PRP-OMPC conjugate is then prepared by direct chemical coupling of both components (coupling is necessary to enhance the immunogenicity of the PRP). After conjugation, the aqueous bulk is adsorbed onto amorphous aluminium hydroxide. The rHBsAg is purified from the recombinant *S. cerevisiae* producer cell media by a combination of chromatographic steps, including ion exchange and hydrophobic interaction chromatography. The purified protein is then formaldehyde-treated and coprecipitated with alum (potassium aluminium sulphate). The individual PRP-OMPC and rHBsAg bulk products are then combined to form Comvax. Each 0.5-ml dose is formulated to contain 7.5 μg PRP conjugated to 125 μg OMPC, 5μg rHBsAg, approximately 225 μg aluminium and 35 μg sodium borate in 0.9% sodium chloride.

Overview of Therapeutic Properties

Before the introduction of *Haemophilus b* conjugate vaccines, Hib was the most frequent cause of bacterial meningitis and a leading cause of serious, systemic bacterial disease in young children globally. Antibodies against PRP have been shown to correlate with protection against Hib disease, and conjugation to the OMPC carrier is believed to enhance the immunological response and immunological memory. The protective efficacy of PRP-OMPC has been demonstrated in clinical trials. Hepatitis B is a significant cause of viral hepatitis. There are 200,000 to 300,000 new cases of hepatitis B infection annually in the U.S., with approximately 1 million Americans being chronic carriers. Recombinant hepatitis B surface antigen has been shown to elicit protective immunity against hepatitis B virus. The combination vaccine Comvax was demonstrated in trials to be highly immunogenic.

Vaccination with Comvax should ideally begin at approximately 2 months of age. It is usually administered in a three-dose regimen (0.5-ml doses at 2, 4, and between 12 and 15 months of age, administered intramuscularly). The product is generally well tolerated. The most frequent side effects are mild, transient signs of inflammation at the injection site, somnolence, and irritability. The product is contraindicated in cases where there is a known hypersensitivity to yeast or to any vaccine component.

Further Reading

Davis, R.L. et al., Impact of the introduction of a combined Hemophilus b conjugate vaccine and hepatitis B recombinant vaccine on vaccine coverage rates in a large west coast health maintainence organization, *Pediatr. Infect. Dis. J.*, 22(7), 657–658, 2003.

Ecokinase (withdrawn from market)

Product Name: Ecokinase (trade name)

Reteplase (international nonproprietary name)

Description: Ecokinase is a recombinant, nonglycosylated, modified form of human tissue plasminogen activator (tPA). It is produced in *Escherichia coli* by recombinant DNA technology and is presented in a lyophilized form to be reconstituted before intravenous administration using a solvent-filled syringe. (See also the monograph for the product Rapilysin, which contains the same active ingredient.)

Approval Date: 1996 (E.U.)

Withdrawal Date: 1999

Therapeutic Indications: Ecokinase was indicated in the thrombolytic therapy of acute myocardial infarction.

Marketing: Marketing authorization was held by Galenus Mannheim, Sandhofer Str. 116, 68305, Mannheim, Germany.

Enbrel

Product Name:	Enbrel (trade name) Etanercept (international nonproprietary name)
Description:	Enbrel is a recombinant fusion protein consisting of the extracellular (ligand-binding portion) of the human tumor necrosis factor (TNF) receptor (p75) and the Fc portion of an IgG antibody. It binds TNF and inhibits the TNF-mediated inflammation process. It is produced by recombinant DNA technology in Chinese hamster ovary (CHO) cells. The final product is supplied in a lyophilized form (25 mg active ingredient per vial) for reconstitution before subcutaneous injection.
Approval Date:	1998 (U.S.); 2000 (E.U.)
Therapeutic Indications:	Enbrel is indicated in the treatment of rheumatoid and some related forms of arthritis.
Manufacturer:	Wyeth Laboratories, New Lane, Havant, Hampshire, P09 2NG, U.K. and Wyeth Medica Ireland, Little Connell, Newbridge, Co. Kildare, Ireland, http://www.wyeth.com (E.U.) Immunex Corporation, 51 University Street, Seattle, WA 98101, http://www.immunex.com (U.S.)
Marketing:	Wyeth Europa Ltd., Huntercombe Lane South, Taplow, Maidenhead, Bershire, SL6 0PH, U.K., http://www.wyeth.com (E.U.) Immunex Corporation, 51 University Street, Seattle, WA 98101, http://www.immunex.com (U.S.)

Manufacturing

Etanercept is a recombinant fusion protein generated by recombinant DNA technology. It consists of the extracellular domain of the TNF-receptor p75 and the Fc of a human IgG1. The gene of the TNF receptor p75 was cloned from human fibroblasts, whereas the Fc region of an IgG1, including the

CH2 and CH3 but not CH1 domains, was cloned using PCR amplification and published nucleotide sequence information. The fusion protein is expressed in a CHO cell line. The product forms a 150-kDa homodimer and is purified from the medium after secretion from the cells. Purification involves several chromatographic steps, ultrafiltration, and procedures for removing viral particles. The final product is presented in a lyophilized form and contains mannitol, sucrose, and trometamol as excipients.

The shelf life of the product is 18 months when stored at 2 to 8°C. Routine evaluative tests are carried out on the final product to ensure its quality and safety.

Overview of Therapeutic Properties

Enbrel can be used for the treatment of rheumatoid arthritis in adults and polyarticular-course juvenile rheumatoid arthritis in children between 4 and 17 years of age when treatment with disease modifying antirheumatic drugs, including methotrexate, have had no adequate response — or in cases of intolerance to those drugs.

In patients with rheumatoid arthritis, TNF accumulates in the joints, stimulating inflammation and destruction of cells. Etanercept, with its binding activity for TNF, blocks the activation of TNF receptors and therefore the TNF-mediated inflammatory process.

Enbrel is administered as a subcutaneous injection twice a week as a 25-mg dose in adults and at a reduced dosage in small children. The presence of the immunoglobulin domain as part of the fusion protein increases the half-life of etanercept in the blood to 70 hours. In clinical trials, Enbrel proved to be effective from within 2 weeks to 1 to 3 months from the first administration. After discontinuation of the treatment, symptoms usually return within 1 month.

Enbrel proved to be safe and well tolerated, with the most common side effects being reaction at the site of injection and infections. Antibodies against Enbrel (although not neutralizing) have been detected and have no consequences upon the effectiveness of the treatment. Antinuclear and anti-ds DNA antibodies have been detected, but no lupus-like syndrome developed. Some blood disorders have been reported, leading to a fatal outcome. Enbrel is contraindicated during pregnancy and lactation.

Further Reading

http://www.enbrel.com
http://www.eudra.org
http://www.fda.gov
http://www.immunex.com
http://www.wyeth.com

Alldred, A., Etanercept in rheumatoid arthritis. *Expert. Opin. Pharmacother.*, 2, 1137–1148, 2001.

Moreland, L.W., Soluble tumor necrosis factor receptor (p75) fusion protein (ENBREL) as a therapy for rheumatoid arthritis, *Rheum. Dis. Clin. North. Am.*, 24, 579–591, 1998.

Engerix-B

Product Name:	Engerix-B (trade name) Hepatitis B vaccine (recombinant) (international nonproprietary name)
Description:	Engerix-B is a vaccine against the hepatitis B virus. It contains the hepatitis B surface antigen (HBsAg) produced by recombinant DNA technology in *Saccharomyces cerevisiae*. Engerix-B is presented as a suspension in vials or in prefilled syringes for intramuscular administration.
Approval Date:	1998 (U.S.)
Therapeutic Indications:	Engerix-B is indicated for immunization against hepatitis B.
Manufacturer:	GlaxoSmithKline Biologicals, Rue de l'institut 89, 1330 Rixensart, Belgium, http://www.gsk.com (U.S.)
Marketing:	GlaxoSmithKline, Research Triangle Park, NC 27709

Manufacturing

Engerix-B is a recombinant vaccine against the hepatitis B virus. It contains the major surface antigen of the hepatitis B virus. This protein is produced in *S. cerevisiae* using recombinant DNA technology. Purification involves several chromatographic steps, including ion exchange and gel permeation chromatography, as well as various filtration and ultracentrifugation procedures. The HBsAg molecules assemble spontaneously into characteristic spherical particles of 20-nm average diameter. The product is adsorbed on aluminium hydroxide. Engerix-B is presented as a suspension containing the HBsAg (active) as well as aluminium hydroxide, sodium chloride, dihydrate disodium phosphate, dihydrate dihydrogen sodium phosphate, and thimerosal as excipients. A formulation for pediatric and adolescent use is supplied without a preservative (thimerosal). Engerix-B is presented as a suspension for intramuscular administration in vials or prefilled syringes.

The shelf life of the product is 36 months when stored at 2 to 8°C in a light-protected container. The quality and safety of the product are validated by extensive testing.

Overview of Therapeutic Properties

Engerix-B is a recombinant vaccine indicated for immunization against the hepatitis B virus. It should be administered intramuscularly into the deltoid muscle in adults and children and into the antero-lateral thigh in infants. The recommended schedule of immunization consists of three injections, with the second injection administered 1 month after the first, and the third injection given 5 months later. In cases where a rapid immunization is required, the third dose may be administered 2 months after the first, with a booster dose 10 months later.

Engerix-B was found to elicit an immune response against the hepatitis B virus. Vaccination against the hepatitis B virus also led to immune protection against hepatitis D, which occurs only in association with hepatitis B, and to a reduction in the incidence of liver cancer.

Engerix-B may be administered with *Haemophilus influenzae* b, hepatitis A, polio, diphtheria, tetanus, pertussis, measles, mumps, rubella, and BCG vaccines.

Engerix-B was found to be safe and well tolerated, with reactions at the site of injection the most commonly reported side effects. Engerix-B should be administered with care during pregnancy and lactation and is contraindicated in the case of severe febrile illness.

Further Reading

http://www.fda.gov

http://www.gsk.com

Andre, F.E. and Safary, A., Summary of clinical findings on Engerix-B, a genetically engineered yeast derived hepatitis B vaccine, *Postgrad. Med. J.* 63(Suppl. 2), 169–177, 1987.

Assad, S. and Francis, A., Over a decade of experience with a yeast recombinant hepatitis B vaccine, *Vaccine,* 18, 57–67, 1999.

Crovari, P. et al., Immunogenicity of a yeast-derived hepatitis B vaccine (Engerix-B) in healthy young adults, *Postgrad. Med. J.,* 63(Suppl. 2), 161–164, 1987.

Goldfarb, J. et al., Comparison study of the immunogenicity and safety of 5- and 10-microgram dosages of a recombinant hepatitis B vaccine in healthy infants, *Pediatr. Infect. Dis. J.,* 15, 764–767, 1996.

Goldfarb, J. et al., Comparison study of the immunogenicity and safety of 5- and 10-microgram dosages of a recombinant hepatitis B vaccine in healthy children, *Pediatr. Infect. Dis. J.,* 15, 768–771, 1996.

Epogen/Procrit

Product Name: Epogen and Procrit (trade names)

Epoetin alfa (international nonproprietary name)

Description: Epogen and Procrit are trade names given to an identical product — human erythropoietin, produced in Chinese hamster ovary (CHO) cells using recombinant DNA technology. The products exhibit an amino acid sequence identical to the naturally occurring molecule but differ somewhat from this in terms of their exact glycosylation pattern. The product stimulates the production of red blood cells and is provided as a liquid formulation for intravenous and subcutaneous administration.

Approval Date: 1989 (Epogen, U.S.); 1990 (Procrit, U.S.)

Therapeutic Indications: Epogen/Procrit is indicated for the treatment of anemia associated with various medical conditions (see Overview of Therapeutic Properties below)

Manufacturer: Amgen Inc., One Amgen Center Drive, Thousand Oaks, CA 91320-1799, http://www. amgen.com

Marketing: Epogen: Amgen Inc., One Amgen Center Drive, Thousand Oaks, CA 91320-1799, http://www.amgen.com
Procrit: Ortho Biotech Products, L.P., Raritan, NJ, http://www.orthobiotech.com

Manufacturing

This product is produced in a CHO cell line using recombinant DNA technology. It is very similar to the naturally occurring molecule and differs only in its glycosylation pattern. Its biological activity appears to remain identical to the native molecule. The glycoprotein is purified from the culture medium using several chromatographic steps and procedures to ensure its microbiological and viral safety. The final product is presented in a liquid form in a number of different formulations to be used for intravenous or subcutaneous administration: A single-dose preparation contains epoetin alfa (active), as well as human albumin, sodium citrate, sodium chloride, and citric acid as excipients. A second single-dose preparation contains a higher concentration of epoetin alfa, as well as human albumin, monohydrate monobasic sodium phosphate, anhydrate dibasic sodium phosphate, sodium citrate, sodium chloride, and citric acid. A multidose preparation is available that contains epoetin alfa, human albumin, sodium citrate, sodium chloride, citric acid, and benzyl alcohol (as a preservative).

Epogen has a shelf life of 24 months when stored at 2 to 8°C. Extensive testing, including SDS-PAGE, Western blotting, isoelectric focusing, HPLC, RIA, *in vivo* bioassays, and microbial and viral analyses, is carried out to ensure the quality and safety of the product.

Overview of Therapeutic Properties

Epoetin alfa is indicated for the treatment of anemia to stimulate the production of red blood cells. Amgen and Ortho Biotech have divided marketing rights in the U.S. Amgen distributes epoetin alfa under the trade name Epogen for the treatment of anemia associated with end-stage renal disease, and Ortho Biotech markets epoetin alfa under the trade name Procrit for all other indications, such as the treatment of anemia associated with renal disease and cancer chemotherapy, anemia and leukemia in HIV-infected patients, and anemia in patients undergoing elective, noncardiac, nonvascular surgery.

Epoetin alfa is generally administered at a dosage suited to the individual patient's needs three times a week, intravenously or subcutaneously, in order to increase the hematocrit level. A maintenance dose, also individually adjusted, is subsequently administered.

Epoetin alfa was found to significantly increase the level of hematocrit, leading to a reduced need for blood transfusions in chronic renal failure patients (whether or not on dialysis), in zidovudine-treated HIV-infected patients, in cancer patients undergoing chemotherapy, and in patients scheduled to undergo elective, noncardiac, nonvascular surgery.

Thrombotic events and an increased mortality were observed in hemodialysis patients who received epoetin alfa, compared to those receiving placebos. Hypertension and seizures were also reported. Epoetin alfa should be administered very carefully during pregnancy, and it is contraindicated during lactation. The multidose formulation containing benzyl alcohol should be avoided for pediatric use.

Further Reading

http://www.amgen.com
http://www.epogen.com
http://www.fda.gov
http://www.orthobiotech.com
http://www.procrit.com
Demetri, G.D. et al., Quality-of-life benefit in chemotherapy patients treated with epoetin alfa is independent of disease response or tumor type: results from a prospective community oncology study. Procrit Study Group, *J. Clin. Oncol.*, 16, 3412–3425, 1998.
Goldberg, M.A., Perioperative epoetin alfa increases red blood cell mass and reduces exposure to transfusions: results of randomized clinical trials, *Semin. Hematol.*, 34 (3 Suppl. 2), 41–47, 1997.
Itri, L.M., The use of epoetin alfa in chemotherapy patients: a consistent profile of efficacy and safety, *Semin. Oncol.*, 29(3 Suppl. 8), 81–87, 2002.
Sullivan, P., Associations of anemia, treatments for anemia, and survival in patients with human immunodeficiency virus infection, *J. Infect. Dis.*, 185(Suppl. 2), 138–142, 2002.
Tang, W.W. et al., Effects of Epoetin alfa on hemostasis in chronic renal failure, *Am. J. Nephrol.*, 18, 263–273, 1998.

Fabrazyme

Product Name: Fabrazyme (trade name)

Agalsidase beta (international nonproprietary name)

Description: Fabrazyme is a recombinant human α-galactosidase A. It is produced by recombinant DNA technology in Chinese hamster ovary (CHO) cells. The 429-amino acid glycoprotein spontaneously dimerizes, yielding the 100-kDa biologically active enzyme. The product is supplied in a lyophilized form (35 mg/vial) to be reconstituted and diluted before use as an intravenous infusion.

Approval Date: 2001 (E.U.)

Therapeutic Indications: Fabrazyme is indicated as a long-term enzyme replacement therapy in patients with Fabry's disease.

Manufacturer: Genzyme Ltd., 37 Hollands Road, Haverhill, Suffolk, U.K., http://www.genzyme.com (manufacturer is responsible for import and batch release in the European Economic Area)

Marketing: Genzyme B.V., Gooimeer 10, 1411 DD Naarden, the Netherlands, http://www.genzyme.com (E.U.)

Manufacturing

The recombinant α-galactosidase is produced by recombinant DNA technology in CHO cells. The purification process includes four chromatographic steps, as well as ultrafiltration and diafiltration steps. The final purified product is presented in a lyophilized form consisting of agalsidase beta (active) and mannitol, sodium phosphate monobasic monohydrate, and sodium phosphate dibasic heptahydrate as excipients.

The shelf life of the product is 18 months when stored at 2 to 8°C. Tests are carried out during processing and on the final product to ensure quality and safety.

Overview of Therapeutic Properties

Fabry's disease is a genetic disease characterized by the deficiency of the lysosomal enzyme α-galactosidase A, leading to an inability to break down certain glycolipids, particularly the glycosphingolipid ceramide trihexoside or globotriaosylceramide (GL-3). The result of the disorder is an accumulation of glycolipids in the walls of vascular cells, particularly in the kidney, heart, and nervous system. To restore enzymatic activity, the treatment of choice is enzyme replacement therapy.

Fabrazyme is administered as a 1 mg/kg dose every 2 weeks as an intravenous infusion. It proved its efficacy in the treatment of Fabry's disease with significant reduction of GL-3 in the vascular endothelium of kidney, skin, and heart tissue.

Due to the rarity of the disease (an estimated 500 to 1000 patients in the E.U.) and the nature of the medicinal product, approval was granted after only preliminary studies, and long-term clinical trials are still ongoing. No other curative treatment is available for patients with Fabry's disease.

Chills and fever were reported as common side effects. Infusion-related reactions have been experienced by half the patients. Hypersensitivity, cardiovascular, and gastrointestinal symptoms have been reported. Most of the patients developed antibodies against Fabrazyme, but no reduction in efficacy was observed. In order to prevent organ damage that could be difficult to reverse, administration of Fabrazyme before the disease develops should be considered.

Fabrazyme should not be administered with chloroquine, amiodarone, benoquin, or gentamicin. No safety and efficacy studies were carried out on patients under 16 or over 65 years old. Fabrazyme is contraindicated during pregnancy and lactation.

Further Reading

http://www.eudra.org
http://www.fabrazyme.com
http://www.genzyme.com
Eng, C.M. et al., A phase 1/2 clinical trial of enzyme replacement in Fabry's disease: pharmacokinetic, substrate clearance, and safety studies, *Am. J. Hum. Genet.*, 68, 711–722, 2001.

Eng, C.M. et al., Safety and efficacy of recombinant human alpha-galactosidase A — replacement therapy in Fabry's disease, *N. Engl. J. Med.*, 345, 9–16, 2001.
Schiffmann R. et al., Enzyme replacement therapy in Fabry's disease: a randomized controlled trial, *J. Am. Med. Assoc.*, 285, 2743–2749, 2001.

Fasturtec/Elitek

Product Name: Fasturtec (trade name E.U.)

Elitek (trade name U.S.)

Rasburicase (international nonproprietary name)

Description: Fasturtec or Elitek is a recombinant form of the *Aspergillus flavus* urate oxidase, the enzyme that converts uric acid to allantoin. It is produced by recombinant DNA technology in yeast cells. The biologically active form of the enzyme is a tetramer, consisting of four identical (34-kDa, 301-amino acid) polypeptides. It is supplied in a lyophilized form (1.5 mg/vial) to be reconstituted and diluted before use as an intravenous infusion.

Approval Date: 2001 (E.U.); 2002 (U.S.)

Therapeutic Indications: Fasturtec is indicated for treatment and prophylaxis of acute hyperuricaemia to prevent acute renal failure in patients with hematological malignancy with a high tumor burden and patients at high risk of a rapid tumor lysis or shrinkage at initiation of chemotherapy.

Manufacturer: Sanofi Winthrop Industrie, 1 rue de l'Abbaye, 76960 Notre Dame de Bondeville, France, http://www.sanofi-synthelabous.com (manufacturer is responsible for batch release in the European Economic Area)

Sanofi-Synthelabo Inc., 9 Great Valley Parkway, P.O. Box 3026, Malvern, PA 19355, http://www.sanofi-synthelabous.com (U.S.)

Marketing: Sanofi-Synthelabo, 174 avenue de France, 75013 Paris, France, http://www.sanofi-synthelabous.com (E.U.)

Sanofi-Synthelabo Inc., 9 Great Valley Parkway, P.O. Box 3026, Malvern, PA 19355, http://www.sanofi-synthelabous.com (U.S.)

Manufacturing

The enzyme rasburicase is produced by recombinant DNA technology. The urate oxidase gene from *A. flavus* is cloned and expressed in a modified *Saccharomyces cerevisiae* strain. The recombinant tetrametric protein is identical to the native form and consists of different isoforms. The purification process involves extraction, concentration, filtration, and several chromatographic steps. The final product is presented in a lyophilized form consisting of rasburicase (active), as well as alanine, mannitol, disodium phosphate, and sodium dihydrogen phosphate as excipients. It is presented with a solvent, used for dilution, containing water for injection and poloxamer 188.

The shelf life of the product is 36 months when stored at 2 to 8°C. Extensive control tests are carried out to ensure quality and safety of the product (including immunoradiometric assays, peptide mapping, activity assays, chromatographies, and SDS-PAGE).

Overview of Therapeutic Properties

Rapid cellular proliferation and turnover associated with malignancy, along with increased rates of cellular lysis due to chemotherapy, results in increased rates of DNA catabolism. This, in turn, leads to a high level of uric acid in the circulation. Humans do not break down uric acid, which is instead removed by the kidney. In the presence of high levels of uric acid, crystals precipitate in the kidney, leading to severe renal failure. The recombinant enzyme Fasturtec is an uridase oxide from *A. flavus* which catalyses the conversion of uric acid into allontoin, a soluble product that is easily excreted from the kidney. Fasturtec is administered as an intravenous infusion daily for 5 to 7 days at a dose of 0.20 mg/kg.

Fasturtec proved its efficacy in controlling uric acid levels in both children and adults. It reduced the level of uric acid significantly more than allopurinol, a product used in conventional treatment. Four hours after administration, the level of uric acid in patients receiving Fasturtec decreased by 85% (reaching normal levels); in the case of allopurinol, the reduction was only 12% after the same time period, and 24 hours were needed for normal levels to be reached. A nonrecombinant form of rasburicase, currently used in treatment, shows higher heterogenicity and lower specific activity than Fasturtec.

Fasturtec is well tolerated and the most common side effects are fever, nausea, and vomiting. Allergic-type reactions have been reported, as well as the production of antirasburicase antibodies. Fasturtec should not be readministered. It should not be administered in case of metabolic disorders such as G6PD deficiency, because the production of H_2O_2 as a product of degradation of uric acid, if not eliminated, could cause hemolytic anemia. Fasturtec is contraindicated during pregnancy and lactation.

Further Reading

http://www.eudra.org

http://www.fda.gov

http://www.sanofi-synthelabous.com

Brant, J.M., Rasburicase: an innovative new treatment for hyperuricemia associated with tumor lysis syndrome, *Clin. J. Oncol. Nurs.*, 6, 12–16, 2002.

Pui, C.H., Rasburicase: a potent uricolytic agent, *Expert Opin. Pharmacother.*, 3, 433–452, 2002.

Pui, C.H. et al., Recombinant urate oxidase for the prophylaxis or treatment of hyperuricemia in patients with leukemia or lymphoma, *J. Clin. Oncol.*, 19, 697–704, 2001.

Forcaltonin

Product Name:	Forcaltonin (trade name) Salmon calcitonin (international nonproprietary name)
Description:	Forcaltonin is a recombinant salmon calcitonin, a 32-amino acid peptide hormone that inhibits bone resorption. It is produced in *Escherichia coli* by recombinant DNA technology and is identical to the chemically synthesized salmon calcitonin. Forcaltonin is supplied as a solution (100 IU activity/ml) for subcutaneous, intramuscular, or intravenous injection.
Approval Date:	1999 (E.U.)
Therapeutic Indications:	Forcaltonin is indicated for the treatment of Paget's disease and hypercalcemia of malignancy.
Manufacturer:	Penn Pharmaceuticals Ltd., 23–24 Tafamaubach Industrial Estate, Tredegar, Gwent NP2 3AA, Wales, U.K., http://www.pennpharm.co.uk (manufacturer is responsible for batch release in the European Economic Area)
Marketing:	Unigene UK Ltd., 63 High Road, Bushey Heath, Herts, WD2 1EE, U.K., http://www.unigene.com

Manufacturing

Salmon calcitonin consists of a 32-amino acid peptide. It is produced in *E. coli* by recombinant DNA technology. The C-terminal residue of native salmon calcitonin is amidated, and this post-translational modification is required for full biological activity. Because *E. coli* is incapabable of carrying out amidation, this post-translational processing is achieved *in vitro* using an α-amidating enzyme obtained from a culture of Chinese hamster ovary (CHO) cells by genetic engineering. The recombinant salmon calcitonin is identical to the chemically synthesized salmon calcitonin. The final purified product is presented as a solution containing salmon calcitonin (active), as well as glacial acetic acid, sodium acetate trihydrate, and sodium chloride as excipients.

The shelf life of the product is 24 months when stored at 2 to 8°C. Tests to ensure quality and safety of the product include RP-HPLC, CEX-HPLC, *in vitro* assay, rat bioassay, and viral safety tests.

Overview of Therapeutic Properties

The overall effect of salmon calcitonin is to increase the amount of calcium and phosphate laid down in bones, with an added effect of significantly reducing serum calcium levels. Forcaltonin is indicated in the treatment of Paget's disease (a bone disease characterized by inflammation and bone deformation) and hypercalcemia of malignancy (tumor associated with increased serum calcium levels). Salmon calcitonin is 30 times more potent in humans than the native human calcitonin. Chemically synthesized calcitonin has been used for more than 20 years (human calcitonin, Cibacalcin, and salmon calcitonin, Calcimar and Miacalcin).

In the case of Paget's disease, Forcaltonin is administered at a dose of 15 μg of calcitonin daily, or at a lower dose of 7.5 μg three times a week, as a subcutaneous or intramuscular injection. Administration normally continues on a weekly basis for a minimum of 3 months. In the case of hypercalcemia of malignancy, Forcaltonin is administered at 15 μg of calcitonin every 6 to 8 hours, or at a higher dose if required.

Forcaltonin proved to be safe and well tolerated, with the most common side effects being nausea, irritation at the injection site, and flushed face and hands. Allergic reactions have been reported and antibodies against Forcaltonin have occasionally developed, though with no apparent effect on efficacy. Forcaltonin should not be administered to children or in cases of persistent hypocalcemia. It is contraindicated during pregnancy and lactation and in patients receiving drugs for heart disorders.

Further Reading

http://www.eudra.org

http://www.unigene.com

Guggi, D. et al., Systemic peptide delivery via the stomach: *in vivo* evaluation of an oral dosage form for salmon calcitonin, *J. Control Release*, 92(1–2), 125–135, 2003.

Lee, D.J. et al., A randomized trial of nasal spray salmon calcitonin in post menopausal Korean women, *Bone*, 32(5), S210, 2003.

Ray, M.V. et al., Production of recombinant salmon calcitonin by *in vitro* amidation of an *Escherichia coli* produced precursor peptide, *Biotechnology*, 11, 64–70, 1993.

Stepan, J.J. and Zikan, V., Calcitonin load test to assess the efficacy of salmon calcitonin. *Clin. Chim. Acta.*, 336(1–2), 49–55, 2003.
Von Tirpitz, C. and Reinshagen, M., Management of osteoporosis in patients with gastrointestinal diseases, *Eur. J. Gastroen. Hepat.*, 15(8), 869–876, 2003.

Forsteo/Forteo

Product Name: Forsteo (trade name E.U.)

Forteo (trade name U.S.)

Teriparatide (international nonproprietary name)

Description: Teriparatide is a recombinant 4,117.8-Da polypeptide identical in amino acid sequence to the N-terminal residues 1-34 of endogenous human parathyroid hormone. Native parathyroid hormone is an 84-amino acid polypeptide that stimulated bone formation by directly affecting bone-forming cells (osteoblasts). Forteo, produced in engineered *Escherichia coli*, induces this same effect. The product is supplied as a sterile solution at a strength of 250 μg active/ml. It is packaged as a 3-ml glass cartridge, preassembled into a disposable pen device for daily subcutaneous administration.

Approval Date: 2002 (U.S.); 2003 (E.U.)

Therapeutic Indications: Treatment of osteoporosis in postmenopausal women.

Manufacturer: Active substance is manufactured by Eli Lilly, Indianapolis, IN, and finished product is manufactured by Lilly France S.A.S., rue du Colonel Lilly, 67640, Fegersheim, France, http://www.lilly.com (E.U. and U.S.)

Marketing: Eli Lilly Netherlands B.V., Grootslag 1-5, NL-3991 RA Houten, the Netherlands (E.U.)
Eli Lilly, Indianapolis, IN, http://www.lilly.com (U.S.)

Manufacturing

Teriparatide is produced in a transformed, nonpathogenic *E. coli* K-12 strain. Following fermentation and product recovery, purification is achieved using a series of chromatographic and filtration steps. Excipients added are mannitol, metacresol, glacial acetic acid, sodium acetate, hydrochloric acid, and sodium hydroxide. After potency and pH adjustment (to pH 4.0), the product is sterile-filtered and aseptically filled into presterile glass cartridges. The cartridges are inserted into a disposable pen device. The product displays a shelf life of 24 months when stored at 4°C.

Overview of Therapeutic Properties

Osteoporosis affects some 75 million people (mainly older women) in Europe, the U.S., and Japan combined. Sufferers display a progressive thinning of the bones (decreased bone mass and microarchitectural deterioration of bone tissue), which leads to bone fragility and an increased risk of bone fracture, particularly of the spine, wrist, and hip.

Human parathyroid hormone (hPTH) is the primary regulator of calcium and phosphate metabolism in bones. It stimulates bone formation by direct effects upon osteoblasts, and it increases intestinal absorption of calcium. These effects are mediated through specific high-affinity cell surface receptors. Teriparatide also binds these receptors and induces a similar biological response to that of native PTH.

Treatment of osteoporosis generally entails a daily injection of Teriparatide (20 μg/80 μl dose) for several months. The treatment normally results in a statistically significant reduction in the frequency of new vertebral fractures, although a similar decrease in the rate of hip fractures was not demonstrated.

Adverse effects associated with Teriparatide administration were usually mild and generally did not trigger discontinuation of therapy. Side effects included dizziness and leg cramps. An increased incidence of osteosarcoma was observed in rats when Teriparatide was administered at levels 3 to 60 times more than the recommended dosage rate for humans. The product, therefore, should not be prescribed for patients who are at increased risk for osteosarcoma, including those with Paget's disease.

Further Reading

http://www.eudra.org
http://www.fda.gov
http://www.lilly.com
Anon., Teriparatide — rhPTH (1-34)-LY 333334 — Forteo ®. Treatment of osteoporosis — parathyroid hormone, *Drug Future*, 26(8), 824–827, 2001.
Anon., Teriparatide (Forteo) for osteoporosis, *Med. Lett. Drugs Ther.*, 45(1149), 9–10, 2003.

Deal, C. and Gideon, J., Recombinant human PTH 1-34 (Forteo): an anabolic drug for osteoporosis, *Clev. Clin. J. Med.*, 70(7), 585–585, 2003.

Wos, J.A. and Lundy, M.W., Patent developments in anabolic agents for treatment of bone diseases, *Expert Opin. Ther. Pat.*, 13(8), 1141–1156, 2003.

Genotropin

Product Name: Genotropin (trade name)

Somatropin (rDNA origin) (common name)

Description: Genotropin is a highly purified preparation of human growth hormone (hGH) produced by recombinant DNA technology. The 191-amino acid single-chain 22.1-kDa polypeptide is identical to native hGH. It is presented as a lyophilized powder in a two-chamber cartridge for use with the Genotropin pen or the Genotropin mixer reconstitution device. One chamber contains the lyophilized powder (available in 1.5, 5.8, or 13.8 mg). The second chamber contains the water for injection-based solvent for product reconstitution. A series of miniquick, single-use, two-chamber cartridges are also available, with the front chamber containing active ingredient levels ranging from 0.2 to 2.0 mg.

Approval Date: 1995 (U.S.)

Therapeutic Indications: Genotropin is indicated for the long-term treatment of pediatric patients who have growth failure due to an inadequate secretion of endogenous growth hormone. Other cases of short stature should be excluded. It is also now approved for long-term replacement therapy in adults with growth hormone deficiency of either childhood or adult onset etiology. Growth hormone deficiency should be confirmed by an appropriate growth hormone stimulation test. It is also indicated for long-term treatment of growth failure in children with the genetic condition Prader-Willi syndrome.

Manufacturer: Pharmacia Pharmaceuticals AB, 181 S-751, Uppsala 1, Sweden

Marketing: Pfizer Inc., 235 East 42nd Street, New York, NY 10017 (U.S.) http://www.pfizer.com

Manufacturing

Genotropin is produced by recombinant DNA technology in an engineered strain of *Escherichia coli*. The producer carries a copy of the native hGH cDNA, placed under an *E. coli* export signal sequence, in a circular plasmid. As a result, the hGH is secreted into the *E. coli* periplasmic space during fermentation, with cleavage of its signal sequence yielding the 191-amino acid product. Fermentation is undertaken under conditions that minimize cell lysis. The outer *E. coli* membrane is then disrupted under conditions that leave the inner cell wall intact. The hGH-containing periplasmic contents are thus released uncontaminated by bacterial cytoplasmic proteins, which renders subsequent downstream processing more straightforward.

Purification entails a combination of chromatographic steps, followed by product formulation, sterile filtration, and lyophilization. Excipients include glycine, sodium phosphate buffer, and mannitol. M-cresol (0.3%) is added to the diluent (water for injection) for multi-use presentations. The product is then extensively tested for purity and identity, utilizing techniques including SDS-PAGE, tryptic peptide mapping, size exclusion chromatography, N- and C- terminal sequence analysis, isoelectric focusing, and HPLC. Immunoassays are used to detect any residual *E. coli* periplasmic proteins, whereas residual *E. coli* DNA is detected by DNA hybridization studies. A rat weight gain assay is used to determine biological potency. The final product is generally assigned a shelf life of 2 years when stored at 2 to 8°C.

Overview of Therapeutic Properties

The weekly dose of Genotropin is generally administered through six or seven separate subcutaneous injections. The dosage level must be adjusted for the individual patient. For pediatric patients, a dosage of 0.16 to 0.24 mg/kg body weight per week is recommended. For adult patients, an initial dosage of 0.04 mg/kg/week is usually given, increasing to a maximum dose of 0.08 mg/kg/week. Treatment of growth hormone-deficient pediatric patients stimulates linear growth and normalizes concentrations of insulin-like growth factor-1 (IGF-1). Treatment of growth hormone-deficient adults results in reduced fat mass, increased lean body mass, alterations in fat metabolism, and normalization of IGF-1 concentrations.

A number of additional Genotropin actions have been demonstrated, including stimulation of skeletal growth and increased muscle cell size and numbers in pediatric patients; increased protein synthesis; and retention of sodium, potassium, and phosphorus.

The product is contraindicated if there is evidence of neoplastic activity. It should be administered with caution to patients with diabetes mellitus because growth hormones may induce insulin resistance. Concomitant glucocorticoid treatment may inhibit the growth-promoting effect of growth hormones. Caution should also be exercised if administering Genotropin to nursing mothers because it is not known if the product is excreted in human milk.

In pediatric clinical trials using Genotropin, adverse effects included injection site reactions, headaches, hypothyroidism, and mild hyperglycemia. Leukemia was reported in a small number of pediatric patients, although a link between it and Genotropin administration has not been established. Some adult patients reported side effects including fluid retention, pain and stiffening of the extremities, and hypoesthesia.

Further Reading

Anon., New indications — Genotropin, *Formulary,* 35(8), 631, 2000.

Darendeliler, F. et al., Follow up height after discontinuation of growth hormone treatment in children with intrauterine growth retardation, *J. Pediatr. Endocr. Met.,* 15(6), 795–800, 2002.

Dorr, H. et al., Are needle-free injections a useful alternative for growth hormone therapy in children? Safety and pharmacokinetics of growth hormone delivered by a new needle-free injection device compared to a fine gauge needle, *J. Pediatr. Endocr. Met.,* 16(3), 383–392, 2003.

Hoybye, C. et al., Growth hormone treatment improves body composition in adults with Prader-Willi syndrome, *Clin. Endocrinol.,* 58(5), 653–661, 2003.

Hoybye, C. et al., The growth hormone — insulin like growth factor axis in adult patients with Prader-Willi syndrome, *Growth Horm. IGF Res.,* 13(5), 269–274, 2003.

GlucaGen

Product Name: GlucaGen (trade name)

Glucagon (rDNA origin) (international nonproprietary name)

Description: GlucaGen is a recombinant hormone identical to naturally occurring human glucagon. The 3500-Da, 29-amino acid single-chain polypeptide is produced in a recombinant *Saccharomyces cerevisiae* strain. GlucaGen is supplied in lyophilized form (1 mg/vial) in a kit that also contains a syringe prefilled with a solution for reconstitution before intravenous, intramuscular, or subcutaneous administration.

Approval Date: 1998 (U.S.)

Therapeutic Indications: GlucaGen is indicated for the treatment of severe hypoglycemia. As a diagnostic aid, it is also used to inhibit motility during radiologic and endoscopic examination of the gastrointestinal tract and for computerized tomography, nuclear magnetic resonance scanning, and digital subtraction angiography.

Manufacturer: Novo Nordisk Pharmaceuticals, Inc., 100 College Road West, Princeton, NJ 08540, http://www.novonordisk-us.com

Marketing: Novo Nordisk Pharmaceuticals, Inc., 100 College Road West, Princeton, NJ 08540, http://www.novonordisk-us.com

Manufacturing

GlucaGen is a recombinant form of human glucagon and is produced in a modified *S. cerevisiae* strain. Manufacture apparently entails upward adjustment of the fermentation media subsequent to upstream processing in order to solubilize the precipitated product. Glucagon is then recovered from the media by a series of filtration and high resolution chromatographic steps. It is identical to the naturally occurring human molecule. GlucaGen is presented in a lyophilized form in a kit with a syringe prefilled with diluent. The final product contains recombinant glucagon hydrochloride (active), as well as lactose, hydrochloric acid, and sodium hydroxide as excipients. After reconstitution, GlucaGen is administered intravenously, intramuscularly, or subcutaneously.

The shelf life of the product is 36 months when stored at 2 to 8°C and 18 months when stored at room temperature between 20 and 25°C. Extensive testing is carried out to ensure the quality and safety of the product.

Overview of Therapeutic Properties

GlucaGen has the same activity as the natural hormone in increasing the blood glucose level by stimulating the conversion of glycogen to glucose and relaxing the smooth muscles of the gastrointestinal tract. GlucaGen is indicated for the treatment of severe hypoglycemia in patients undergoing treatment with insulin. It is also used as a diagnostic aid during gastrointestinal radiologic and endoscopic examination and during computerized tomography, nuclear magnetic resonance scanning, and digital subtraction angiography. Anticholinergic agents may be used to relax the smooth muscles in the gastrointestinal tract, but they lead to increased side effects compared to GlucaGen.

GlucaGen is administered intravenously, intramuscularly, or subcutaneously at the dosage required for the diagnostic examination or, in the case of the treatment of hypoglycemia, according to body weight.

GlucaGen has proven to be safe and is associated only with mild side effects, including vomiting, nausea, allergic reactions, and tachycardia. Caution should be taken when administering GlucaGen to patients with insulinoma and glucagonoma. GlucaGen is contraindicated in patients with pheochromocytoma.

Further Reading

http://www.novonordisk-us.com

Anon., GlucaGen — recombinant glucagon for severe HTN and as a radiologic diagnostic aid, *Formulary*, 33(8), 710–710, 1998.

Fonjallaz, P. and Loumaye, E., Glucagon rDNA origin (GlucaGen R) and recombinant LH, *J. Biotechnol.*, 79(2), 185–189, 2000.

Harris, G. et al., Glucagen ® administration — underevaluated and undertaught, *Diabetologia*, 40 (Suppl.), 47, 1997.

Glucagon

Product Name: Glucagon (trade name)

Glucagon (rDNA origin) (international nonproprietary name)

Description: Glucagon is a recombinant glucagon produced in a modified *Escherichia coli* strain and is chemically identical to the native human hormone. The 29-amino acid single-chain polypeptide is presented as a lyophilized powder (1 mg/vial) to be reconstituted before intravenous, intramuscular, or subcutaneous administration. Glucagon is supplied in a Glucagon emergency kit for low blood sugar and in a Glucagon diagnostic kit, which contain the same amount of hormone and are provided with a syringe prefilled with diluent.

Approval Date: 1998 (U.S.)

Therapeutic Indications: Glucagon is indicated for the treatment of severe hypoglycemia and for use as a diagnostic aid in radiologic examination of the stomach, duodenum, small bowel, and colon in cases where diminished intestinal motility would be advantageous.

Manufacturer: Eli Lilly and Co., Lilly Corporate Center, Indianapolis, IN 46285, http://www.lilly.com

Marketing: Eli Lilly and Co., Lilly Corporate Center, Indianapolis, IN 46285, http://www.lilly.com

Manufacturing

Glucagon is a recombinant form of human glucagon. It is produced in a modified *E. coli* strain that has been transformed with the human glucagon-encoding gene. Glucagon is extracted from the bacterial culture and purified using a combination of filtration and several high-resolution chromatographic steps. It is presented as a lyophilized powder in two packages, of identical content —

a Glucagon emergency kit for low blood sugar and a Glucagon diagnostic kit. Each package contains a powder consisting of recombinant glucagon hydrochloride, lactose and hydrochloric acid (excipients), and a syringe prefilled with the diluting solution of glycerin and hydrochloric acid (Hyporet). After reconstitution, Glucagon is administered intravenously, intramuscularly, or subcutaneously.

The shelf life of the product is 18 months when stored at 20 to 25°C. Tests are carried out to ensure the quality and safety of the product.

Overview of Therapeutic Properties

Glucagon retains the biological activity of the natural molecule in increasing the level of glucose in the blood and relaxing the smooth muscles of the gastrointestinal tract. Glucagon is indicated for the treatment of severe hypoglycemia and as a diagnostic aid in the radiologic examination of the stomach, duodenum, small bowel, and colon in situations where diminished intestinal motility is required. Anticholinergic agents are as effective as Glucagon in diminishing intestinal motility but have the disadvantage of producing greater numbers of side effects.

Glucagon should be administered intravenously, intramuscularly, or subcutaneously at a dosage based on body weight, in the case of hypoglycemia, or dependent on the onset and duration required for radiologic examination.

Allergic reactions have been reported after Glucagon administration. Glucagon is contraindicated in patients with insulinoma and pheochromocytoma.

Further Reading

http://www.fda.gov
http://www.lilly.com
See also the GlucaGen monograph and its associated further reading.

Gonal-F

Product Name:	Gonal-F (trade name)
	Follitropin alfa (international nonproprietary name)
Description:	Follitropin alfa is a recombinant human follicle stimulating hormone (r-hFSH) produced in mammalian cells by recombinant DNA technology. It is supplied in a lyophilized form at various strengths (37.5–150 IU/vial) and is reconstituted before administration as subcutaneous injection.
Approval Date:	1995 (E.U.), 1997 (U.S.)
Therapeutic Indications:	Gonal-F is indicated in patients undergoing superovulation for assisted reproductive technologies, such as *in vitro* fertilization (IVF), gamete intrafallopian transfer (GIFT), and zygote intrafallopian transfer (ZIFT). Gonal-F is indicated for anovulation in women who have had no adequate response with clomiphene citrate. Gonal-F is also indicated for the stimulation of spermatogenesis concomitant with human chorionic gonadotrophin therapy in men with congenital or acquired hypogonadotrophic hypogonadism.
Manufacturer:	Serono Pharma S.p.A., Zona Industriale di Modugno, 70123 Bari, Italy, http://www.serono.com
Marketing:	Ares-Serono (Europe) Ltd., 24 Gilbert Street, London W1Y 1RJ, U.K., http://www.serono.com (E.U.) Serono Laboratories, Inc., One Technology Place, Rockland, MA 02370, http://www.serono.com (U.S.)

Manufacturing

r-hFSH consists of two subunits, α (92 amino acids) and β (111 amino acids), and is identical in sequence to the native human follicle stimulating hormone (FSH). The genes encoding the subunits have been cloned into an appropriate plasmid and transfected into Chinese hamster ovary (CHO) cells for production of the recombinant protein. The purification process includes multiple high resolution chromatographic steps and ultrafiltration, as well as additional viral inactivation steps. The final product is presented in a lyophilized form containing follitropin alfa (active), as well as sucrose, sodium dihydrogen phosphate monohydrate, disodium phosphate dihydrate, phosphoric acid, and sodium hydroxide as excipients.

The shelf life of the product is 24 months when stored at 2 to 8°C. Routine evaluative tests are carried out to ensure the quality and safety of the product.

Overview of Therapeutic Properties

FSH regulates the growth of ovarian follicles in females and the induction of spermatogenesis in males and is indicated in infertility treatment for both women and men. Gonal-F is used to stimulate ovarian follicular growth in infertile female patients who do not ovulate and in whom anovulation is related to hormonal imbalance and when no response was achieved with clomiphene citrate. Gonal-F is also used to induce multifollicular development in female patients undergoing assisted reproduction technologies such as IVF, GIFT, and ZIFT. Gonal-F is also used in combination with human chorionic gonadotrophin to stimulate spermatogenesis in male patients with hypogonadotropic hypogonadism when infertility is not the result of primary testicular failure.

Gonal-F is administered as a subcutaneous injection at a dose that can be adjusted for individual patients. Self-administration continues on a daily basis for females and three times a day for males for a number of weeks or months, according to the treatment.

The r-hFSH proved to be effective in the treatment of infertility, showing the same results as urinary FSH (human FSH extracted from female urine). Gonal-F offers the advantage (over urinary FSH) of a purer product, with the urinary FSH preparation containing detectable LH.

Gonal-F proved to be safe in the treatment of female and male infertility. Local reactions at the site of injection were the most common side effects. Ovarian hyperstimulation syndrome was the most relevant side effect in females, leading to the development of cysts. In men, the most common side effects were breast pain, fatigue, and breast enlargement. Gonal-F is contraindicated during pregnancy and lactation.

Further Reading

http://www.eudra.org
http://www.fda.gov
http://www.serono.com

Bergh, C. et al., Recombinant human follicle stimulating hormone (r-hFSH; Gonal-F) versus highly purified urinary FSH (Metrodin HP): results of a randomized comparative study in women undergoing assisted reproductive techniques, *Hum. Reprod.*, 12, 2133–2139, 1997.

Liu, P.Y. et al., Efficacy and safety of recombinant human follicle stimulating hormone (Gonal-F) with urinary human chorionic gonadotrophin for induction of spermatogenesis and fertility in gonadotrophin-deficient men, *Hum. Reprod.*, 14, 1540–1545, 1999.

Prevost, R.R., Recombinant follicle-stimulating hormone: new biotechnology for infertility, *Pharmacotherapy*, 18, 1001–1010, 1998.

HBVAXPRO

Product Name:	HBVAXPRO (trade name)
	Hepatitis B vaccine, recombinant (common name)
Description:	HBVAXPRO is a recombinant vaccine against hepatitis B. It is a new formulation of recombinant hepatitis B vaccine without thiomersal. HBVAXPRO consists of the 24-kDa major surface antigen of the hepatitis B virus obtained by recombinant DNA technology. HBVAXPRO is presented in a prefilled syringe as a suspension for intramuscular injection.
Approval Date:	2001 (E.U.)
Therapeutic Indications:	HBVAXPRO is indicated for active immunization against hepatitis B in children, adolescents, adults, and dialysis patients.
Manufacturer:	Merck Sharp and Dohme, B.V., Waarderweg 39, 2003 PC Haarlem, the Netherlands, http://www.merck.com (manufacturer is responsible for batch release in the European Economic Area)
Marketing:	Aventis Pasteur MSD SNC, 8 rue Jonas Salk, 69007 Lyon, France, http://www.aventispasteur.com

Manufacturing

The recombinant hepatitis B vaccine HBVAXPRO consists of the major membrane protein, the S protein, of the hepatitis B virus. It is produced in a genetically modified *Saccharomyces cerevisiae* strain by recombinant DNA technology. After purification involving several chromatographic steps, the protein is treated with formaldehyde and adsorbed on aluminum hydroxyphosphate sulfate. The final product consists of the antigenic protein, sodium chloride, and borax. No thiomersal, a preservative that contains mercury, is added to the final product. HBVAXPRO is presented as a suspension for intramuscular injection.

The shelf life of the product is 36 months when stored at 2 to 8°C. Tests for sterility, identity, potency, pH, and for the presence of contaminants are carried out to ensure the quality and safety of the vaccine.

Overview of Therapeutic Properties

HBVAXPRO is indicated for immunization against hepatitis B. It is administrated as a three-dose vaccine, with 1 month between the first and the second injection, and a 6-month interval between the first and the third, or, for rapid immunity, as three doses 1 month apart and a fourth dose after 1 year. Different formulations are offered for different categories of patients: 5 μg in 0.5 ml for children and adolescents (up to 15 years old), 10 μg in 1 ml for adolescents (15 years or older) and adults, and 40 μg in 1 ml for predialysis and dialysis patients. A special dosage schedule is indicated in the case of exposure to the hepatitis B virus. A booster vaccination may be included in some vaccination schedules.

Clinical trials to establish safety and immunogenicity were carried out on the formulation containing thiomersal. The thiomersal-free vaccine has been tested for safety and efficacy in preclinical trials; immunological protection against the hepatitis B virus and safety have been extrapolated either from other thiomersal-free vaccines studies or from studies on particular categories of patients. Side effects commonly reported after administration of vaccines include reactions at the injection site, headache, nausea, vomiting, fever, and fatigue. Severe side effects rarely have been reported. Studies on other thiomersal-free vaccines showed less immunogenicity than the same vaccines containing thiomersal, but this minor reduction in immunogenicity is not considered of clinical relevance. HBVAXPRO should not be administered in the case of fever.

Further Reading

http://www.aventispasteur.com
http://www.eudra.org
http://www.fda.gov
http://www.merck.com

Helixate NexGen/Helixate FS/Kogenate FS

Product Name: Helixate NexGen (trade name E.U.)

Helixate FS (trade name U.S.)

Kogenate FS (trade name U.S.)

Octocog alfa (international nonproprietary name)

Description: Octocog alfa is a recombinant human coagulation factor VIII. The glycoprotein is produced by recombinant DNA technology in baby hamster kidney (BHK) cells and is supplied in a lyophilized form at various strengths (250, 500, and 1000 IU/vial). It is reconstituted before intravenous injection.

Approval Date: 2000 (E. U. and U.S.). An alternative formulation, with human albumin as a stabilizer, was approved in 1993 in the U.S.

Therapeutic Indications: The product is indicated for treatment and prophylaxis of bleeding in patients with hemophilia A. (The preparation does not contain von Willebrand's factor and therefore is not indicated in von Willebrand's disease.)

Manufacturer: Bayer AG, 51368 Leverkusen, Germany, http://www.pharma.bayer.com (manufacturer is responsible for import and batch release in the European Economic Area)
Bayer Corporation, 800 Dwight Way, P.O. Box 1986, Berkeley, CA 94701-1986, http://www.pharma.bayer.com (U.S.)

Marketing: Bayer AG, 51368 Leverkusen, Germany, http://www.pharma.bayer.com (E.U.)
Bayer Corporation, 800 Dwight Way, P.O. Box 1986, Berkeley, CA 94701-1986, http://www.pharma.bayer.com (for marketing in U.S. with the trade name Kogenate FS)
Aventis Behring, 1020 First Avenue, King of Prussia, PA 19406, http://www.aventisbehring.com (for marketing in U.S. with the trade name Helixate FS)

Manufacturing

The recombinant coagulation factor VIII is produced by recombinant DNA technology. The human coagulation factor VIII gene is expressed in BHK cells, with post-translational modifications largely similar to the plasma-derived factor VIII being observed. Helixate NexGen differs from the previously approved Helixate in the purification process and in the use of sucrose instead of human albumin as a stabilizer. The purification process involves several chromatographic steps, including immunoaffinity chromatography using a murine IgG, as well as ultrafiltration. A solvent/detergent treatment step is included as a viral inactivation measure. The final product is presented in a lyophilized form consisting of octocog alfa (active), as well as glycine, sodium chloride, calcium chloride, histidine, and sucrose as excipients.

The shelf life of the product is 23 months when stored at 2 to 8° C. Control tests are carried out to ensure quality and safety of the product.

Overview of Therapeutic Properties

Helixate NexGen is indicated for the treatment and prophylaxis of patients with hemophilia A. Hemophilia A is a genetic disease characterized by a lack of, or low levels of, coagulation factor VIII, which is involved in the cascade for the formation of a fibrin clot. Lack of coagulation factor VIII results in profused bleeding, occurring spontaneously or after trauma or surgery.

Helixate NexGen is administered as an intravenous injection at the particular dosage required for each patient. Helixate NexGen proved to be as effective as the plasma-derived product, with increased viral safety, and as effective as the previously approved product Helixate. Helixate NexGen is well tolerated and safe. The most common side effects were reactions at the injection site, hypersensitivity, and an unusual taste. A higher dosage should be utilized in the presence of inhibitors of coagulation factor VIII.

Further Reading

http://www.aventisbehring.com
http://www.eudra.org
http://www.fda.gov
http://www.kogenatefs-usa.com
http://www.pharma.bayer.com

Abshire, T.C. et al., Sucrose formulated recombinant human antihemophilic factor VIII is safe and efficacious for treatment of hemophilia A in home therapy — International Kogenate-FS Study Group, *Thromb. Haemost.*, 83, 811–816, 2000.

Boedeker, B.G., Production processes of licensed recombinant factor VIII preparations, *Semin. Thromb. Hemost.*, 27, 385–394, 2001.

Lusher, J.M. et al., Recombinant factor VIII for the treatment of previously untreated patients with hemophilia A. Safety, efficacy, and development of inhibitors. Kogenate Previously Untreated Patient Study Group, *N. Engl. J. Med.*, 328, 453–459, 1993.

Hepacare (withdrawn from market)

Product Name:	Hepacare (trade name)
	Triple antigen hepatitis B vaccine, recombinant (international nonproprietary name)
Description:	Hepacare is a recombinant hepatitis B vaccine consisting of purified pre-S1, pre-S2, and S surface hepatitis B antigens. It is produced by recombinant DNA technology in murine cells and is supplied in a prefilled syringe as a suspension (20 μg/ml) for intramuscular injection.
Approval Date:	2000 (E.U.)
Withdrawal Date:	2002 (apparently for commercial reasons)
Therapeutic Indications:	Hepacare was indicated for active immunization against the hepatitis B virus infection in nonimmune adults.
Manufacturer:	Medeva Pharma Limited, Gaskill Road, Speke, Liverpool, L29 9GR, U.K., http://www.medeva.co.uk (now Celltech Pharmaceuticals Ltd.)
Marketing Authorization Holder:	Medeva Pharma Limited, Regent Park, Kingston Road, Leatherhead, Surrey KT22 7PQ, U.K., http://www.medeva.co.uk (now Celltech Pharmaceuticals Ltd.)

Manufacturing

The recombinant triple antigen hepatitis B vaccine is produced by recombinant DNA technology. The genes for the antigens are cloned and expressed in murine cells. The viral particles are identical to the natural antigens, consisting of glycosylated and nonglycosylated particles.

The purification process includes ultrafiltration steps and procedures to inactivate and remove viruses. The final purified product is presented as a suspension consisting of pre-S1, pre-S2, and S surface antigens, along with hydrated aluminum oxide and sodium chloride as excipients. The suspension is not buffered in order to obtain better efficacy. The shelf life of the product is 24 months when stored at 2 to 8°C. Extensive control tests are carried out to ensure quality and safety of the product including SDS-PAGE, IEF, CD, SEC, rp-HPLC, immunoassays, Western blots, deglycosylation studies, *in situ* detection, tests for bio burden, mycoplasma, and viruses.

Overview of Therapeutic Properties

Hepatitis B is a chronic disease with a very high incidence worldwide. It is reported that 1% of infected adults die from hepatic failure, while 5 to 10% of infected adults and 80 to 90% of infected children are chronic carriers who could, at any stage, develop liver disorders. The carries also serve as the reservoirs of infection that are responsible for the more than 50 million people infected each year. No satisfactory treatment for the disease is available, and vaccination remains the prophylactic treatment of choice. The first generation of vaccines against hepatitis B were derived from the plasma of infected subjects; the second generation (conventional vaccines currently available on the market) are recombinant vaccines consisting of the surface S antigen of the hepatitis virus. Hepacare is a third-generation vaccine consisting of the antigen S, as well as two more surface antigens, pre-S1 and pre-S2, which proved to be immunogenic. Hepacare is a recombinant product produced by genetic engineering. Different antigen subtypes have been found, but immunization against one form provides protection against all subtypes.

Hepacare is administered intramuscularly as a two-dose vaccination, 20 μg each at a 1-month interval in adults with a normal immune response. It is administered as a three-dose vaccination, with the third dose 6 months after the first in adults with a suboptimal immunoresponse (males, smokers, and individuals who are obese or older than 40). Studies showed that 96% of adults receiving the two-dose regimen and the three-dose regimen developed antibodies against hepatitis after 6 months and 7 months, respectively.

Hepacare induces better immunoprotection (higher titres) than conventional vaccines, induces immunoresponses in subjects who do not respond to conventional vaccines, and the two-dose immunization offers rapid protection (compared to the 6 months with conventional vaccines) for adults at risk.

Vaccination against hepatitis with conventional vaccines generally offers immune protection for 5 years. High titres have been observed in adults after 24 months from the two-dose administration of Hepacare, and long-term studies are ongoing.

Hepacare is as well tolerated as other hepatitis vaccines. The most common side effect is pain at the site of injection, probably due to the low pH of the solution. Hepacare should not be administered in cases of fever.

Further Reading

http://www.eudra.org
http://www.medeva.co.uk
McDermott, A.B. et al., Hepatitis B third-generation vaccines: improved response and conventional vaccine non-response — evidence for genetic basis in humans, *J. Viral Hepat.*, 5, S2, 9–11, 1998.
Page, M. et al., A novel, recombinant triple antigen hepatitis B vaccine (Hepacare), *Intervirology*, 44, 88–97, 2001.
Young, M.D. et al., Adult hepatitis B vaccination using a novel triple antigen recombinant vaccine, *Hepatology*, 34, 372–376, 2001.

Herceptin

Product Name: Herceptin (trade name)

Trastuzumab (international nonproprietary name)

Description: Trastuzumab is a humanized monoclonal antibody with binding activity for the human epidermal growth factor receptor 2, HER2, which is found on the surface of metastatic breast cancer cells. Herceptin is expressed in mammalian cells and is supplied in a lyophilized form (150 mg/vial) to be reconstituted and diluted before administration as an infusion.

Approval Date: 1998 (U.S.); 2000 (E.U.)

Therapeutic Indications: Herceptin is indicated for the treatment of patients with metastatic breast cancer whose tumors overexpress HER2.

Manufacturer: Hoffmann-La Roche AG, Emil-Barell-Str 1, 79639 Grenzach-Wyhlen, Germany, http://www.roche.com (manufacturer is responsible for import and batch release in the European Economic Area)
Genentech, Inc., 1 DNA Way, South San Francisco, CA 94080-4990, http://www.gene.com (U.S.)

Marketing: Roche Registration Limited, 40 Broadwater Road, Welwyn Garden City, Hertfordshire, AL7 3AY, U.K., http://www.roche.com (E.U.)
Genentech, Inc., 1 DNA Way, South San Francisco, CA 94080-4990, http://www.gene.com (U.S.)

Manufacturing

The anti-HER2 murine monoclonal antibody was generated by hybridoma technology following immunization of mice with cells expressing the HER2 protein on their surface and with membranes containing the HER2 protein. The murine monoclonal antibody was then humanized by grafting its complementarity-determining regions onto a human IgG1 kappa framework, resulting in an antibody, trastuzumab, with a three-fold increased affinity for HER2 compared to the murine parent antibody. The antibody is produced in Chinese hamster ovary (CHO) cells in a serum-free medium. Purification involves several chromatographic steps, including affinity chromatography (using Protein A), cation and anion ion exchanges, and hydrophobic interaction chromatography. The purification process inactivates and removes DNA, viruses, and endotoxins. The final product is presented in a lyophilized form containing the antibody trastuzumab (active), as well as L-histidine hydrochloride, trehalose dihydrate, and polysorbate 20 as excipients.

The shelf life of the product is 24 months when stored at 2 to 8°C. Routine evaluative tests are carried out on the final product to ensure quality and safety of the product.

Overview of Therapeutic Properties

About 10% of breast cancers develop into metastatic breast cancers. The life expectancy of patients is 6 to 7 years but decreases to 10 to 12 months from diagnosis in the case of metastatic breast cancers that overexpress the HER2 protein. The HER2 protein, overexpressed in 25 to 30% of metastatic breast cancers, is encoded by the c-erb 2 proto-oncogene and is involved in control of the cell proliferation. When the oncogene is activated, the HER2 protein is overexpressed and induces abnormal proliferation of cells. Herceptin is indicated for the treatment of metastatic breast cancer only when HER2 is overexpressed. The monoclonal antibody binds specifically to the tumor cells overexpressing HER2, thus inhibiting cell proliferation and inducing antibody-directed cell-mediated cytotoxicity. This results in reducing metastasis while not affecting normal cells, therefore limiting side effects.

Herceptin is indicated in combination with paclitaxel in the treatment of patients who have not received chemotherapy for their metastatic disease, or where anthracycline was not suitable. It is also indicated in the treatment of patients who did not respond to at least two chemotherapy treatments, including anthracycline and taxane and, where applicable, hormonal therapy for their metastatic disease.

Herceptin is administered by intravenous infusion, first as a loading dose of 4 mg/kg and then weekly as a maintenance dose of 2 mg/kg until the disease regresses.

Studies on patients undergoing their second chemotherapy regimen with paclitaxel showed an increased tumor response and an increased duration of the response, as well as a 3-month-longer survival period when Herceptin was part of the treatment.

Mild infusion-related reactions have been reported during the first administration of Herceptin. Hypersensitivity reactions that are rarely severe and pulmonary complications have been observed, some with fatal outcomes, particularly in patients with dyspnea at rest. Herceptin should not be administered in combination with anthracycline because of possible heart failure. Herceptin is contraindicated during pregnancy and lactation.

Further Reading

http://www.eudra.org
http://www.fda.gov
http://www.gene.com
http://www.herceptin.com
http://www.roche.com
Leyland-Jones, B. and Smith, I., Role of Herceptin in primary breast cancer: views from North America and Europe, *Oncology*, 61(Suppl. 2), 83–91, 2001.
Pegram, M.D. et al., Phase II study of receptor-enhanced chemosensitivity using recombinant humanized anti-p185HER2/neu monoclonal antibody plus cisplatin in patients with HER2/neu-overexpressing metastatic breast cancer refractory to chemotherapy treatment, *J. Clin. Oncol.*, 16, 2659–2671, 1998.

Hexavac

Product Name: Hexavac (trade name)

Diphtheria, tetanus, acellular pertussis, hepatitis B recombinant, inactivated poliomyelitis, conjugated *Haemophilus influenzae* type b (Hib) vaccine, adjuvanted (common name)

Description: Hexavac is a combined vaccine against diphtheria, tetanus, pertussis, hepatitis B, poliomyelitis, and Hib. It is a new, liquid formulation that combines vaccines already approved in the E.U. Hexavac consists of diphtheria and tetanus toxoids, two pertussis antigens (pertussis toxoid and filamentous haemagglutinin), the major surface antigen of the hepatitis B virus obtained by recombinant DNA technology, inactivated polioviruses (types 1, 2, and 3), and Hib polysaccharide. Hexavac is presented in a prefilled syringe as a suspension for intramuscular injection.

Approval Date: 2000 (E.U.)

Therapeutic Indications: Hexavac is indicated for primary vaccination in infants and booster vaccination in toddlers against diphtheria, tetanus, pertussis, hepatitis B, poliomyelitis, and Hib.

Manufacturer: Aventis Pasteur SA, Campus Mérieux, 1541 Avenue Marcel Mérieux, 69280 Marcy l'Etoile, France, http://www.aventispasteur.com (manufacturer is responsible for batch release in the European Economic Area)

Marketing: Aventis Pasteur MSD SNC, 8 rue Jonas Salk, 69007 Lyon, France, http://www.aventispasteur.com

Manufacturing

The antigenic components of Hexavac are produced and purified separately and combined into the final product. Diphtheria toxoid is obtained from a culture of *Corynebacterium diphtheriae*, detoxified using formalin, and purified by salt precipitation. Tetanus toxoid is obtained from a *Clostridium tetani* culture, detoxified with formalin, and purified by ammonium sulphate precipitation. Pertussis antigens, pertussis toxin, and filamentous hemagglutinin are obtained from a culture of *Bordetella pertussis*; pertussis toxin is treated with glutaraldehyde for detoxification, while filamentous hemagglutinin is used in its native form. The surface hepatitis B antigen is produced in a genetically modified *Saccharomyces cerevisiae* strain by recombinant DNA technology. Purification includes chromatographic steps for removal of yeast proteins, denaturation with formaldehyde, and co precipitation with aluminum. IPV types 1, 2, and 3 are obtained from a Vero cell culture and purified using chromatography and ultrafiltration; polioviruses are inactivated using formaldehyde. Hib polysaccharide is obtained from *H. influenzae*. It is purified, activated with cyanogen bromide (to bind adipic dihydrazide acid), and conjugated to tetanus toxoid.

Hexavac consists of antigenic components and aluminium hydroxide (as adjuvant), disodium phosphate, monopotassium phosphate, sodium carbonate, sodium bicarbonate, trometamol, sucrose, medium 199 (containing mainly amino acids, mineral salts, and vitamins), and traces of the antibiotics neomycin, streptomycin and polymyxin B. It is presented as a liquid formulation for intramuscular injection. The shelf life of the product is 24 months when stored at 2 to 8°C. Extensive tests are carried out to ensure the quality, potency, and safety of the vaccine.

Overview of Therapeutic Properties

The combined vaccine Hexavac is indicated for primary and booster vaccination of infants and toddlers, respectively, to induce immunoprotection against diphtheria, tetanus, pertussis, hepatitis B, poliomyelitis viruses, and Hib. Hexavac should be administered as an intramuscular injection as a 3-dose vaccine in infants, with a 1-month interval between administrations. The booster vaccine is administered 6 to 14 months after completion of the primary vaccination.

Hexavac proved as effective as other combined vaccines in inducing immunoprotection against diphtheria, tetanus, pertussis, poliomyelitis, hepatitis B, and Hib. As already found in other combined vaccines, the antibody titre of anti-Hib antibodies was lower than if the vaccine was administered separately. Nevertheless, the vaccine offered the same level of immunoprotection as when administered separately, through the involvement of a memory response.

Hexavac is safe and well tolerated with side effects comparable to other combined vaccines, except for a local higher reactogenicity (redness). Reactogenicity was slightly higher after a booster vaccination compared to a primary vaccination, as found for other vaccines.

Caution should be exercised in infants with thrombocytopenia or a bleeding disorder, and the vaccine should not be administered in the case of fever.

Further reading

http://www.aventispasteur.com

http://www.eudra.org

Liese, J.G. et al., Large scale safety study of a liquid hexavalent vaccine (D-T-acP-IPV-PRP—T-HBs) administered at 2, 4, 6 and 12-14 months of age, *Vaccine*, 20, 448–454, 2001.

Mallet, E. et al., Immunogenicity and safety of a new liquid hexavalent combined vaccine compared with separate administration of reference licensed vaccines in infants, *Pediatr. Infect. Dis. J.*, 19, 1119–1127, 2000.

Humalog

Product Name: Humalog (trade name)

Insulin lispro (international nonproprietary name)

Description: Humalog is a fast-acting insulin analogue . It is chemically identical to human insulin, except for an inversion of the natural proline-lysine sequence at positions 28 and 29 of the insulin B chain. It is produced by recombinant DNA technology in *Escherichia coli*. Its final product presentation is as a solution for injection, containing Zn-insulin lispro crystals, and it is administered subcutaneously. A new formulation consisting of a lesser amount of Humalog dissolved in water and the remainder in suspension with protamine sulphate, or completely in suspension with protamine sulphate, is also available.

Approval Date: 1996 (U.S. and E.U.);
1998 for Humalog Mix 25, Humalog Mix 50, and Humalog NPL (E.U.);
1999 for Humalog 75/25 and Humalog 50/50 (U.S.)

Therapeutic Indications: Humalog is used to promote normal glucose homeostasis in patients suffering from diabetes mellitus. It can also be used for the initial stabilization of the disease.

Manufacturer: Lilly France S.A., Rue du Colonel Lilly, 67640 Fegersheim, France; Lilly Pharma Fertigung und Distribution GmbH and Co. KG, Teichweg 3, 35396 Giessen, Germany; and Lilly S.A., Avda. de la Industria 30, 28108 Alcobendas (Madrid), Spain, http://www.lilly.com (manufacturer is responsible for import and batch release in the European Economic Area)
Lilly France S.A., Rue du Colonel Lilly, 67640 Fegersheim, France (U.S.)

Marketing: Eli Lilly Nederland B.V., Krijtwal 17-23, 3432 ZT Nieuwegein, the Netherlands (E.U.)
Eli Lilly & Co., Indianopolis, IN 46285, http://www.lilly.com (U.S.)

Manufacturing

The production process is initiated by large-scale fermentation of a laboratory strain of *E. coli* (K-12) harboring a plasmid that contains a gene encoding LysB28-ProB29 human proinsulin. The gene product accumulates intracellularly. Upon completion of the fermentation step, the *E. coli* cells are harvested and lysed. The crude proinsulin extract is treated with trypsin and carboxypeptidase B, thereby liberating mature insulin lispro. The latter is purified to homogeneity by a combination of chromatographic steps and is finally crystallized in the presence of zinc. It is formulated as a solution for injections which, in addition to insulin lispro, contains the following excipients: WFI (vehicle), M-cresol and phenol (as preservative and stabilizer in some formulations), glycerol (tonicity modifier), zinc oxide (stabilizer), dibasic sodium phosphate (buffer), sodium hydroxide, and hydrochloric acid (adjustment of final pH to 7.0 to 7.8). Humalog may also be presented in a different formulation containing 25 or 50% of Humalog dissolved in water and the rest as a suspension with protamine sulphate — or completely as a suspension with protamine sulphate. Humalog in suspension with protamine sulphate has a prolonged action. Humalog is available as a liquid formulation in vials for injections or with a pen device for the administration of several doses.

The product displays a shelf life of 24 months when stored at 2 to 8°C. Routine evaluative tests carried out on the final product by the manufacturer include identity by rabbit hypoglycemia assay, potency by RP-HPLC, purity and identity by RP-HPLC and SE-HPLC, detection of *Escherichia* host protein contaminants by immunoassay, detection of proinsulin and C-peptide contaminants by immunoassay, and endotoxin and pyrogen tests.

Overview of Therapeutic Properties

The therapeutic effect (ability to lower blood glucose levels) of unmodified human insulin and insulin lispro is essentially the same. In the case of insulin lispro, however, a faster rate of absorption from the site of administration is observed, with an associated more rapid onset of action. Peak serum insulin levels usually occur 120 to 240 minutes after subcutaneous administration of unmodified insulin. SC administration of insulin lispro is followed by peak-serum levels observed within 30 to 120 minutes. Regular human insulin is best administered 30 to 60 minutes before a meal. This can often be inconvenient. Insulin lispro, however, is administered 15 minutes before a meal or immediately after a meal, so forward planning of exact meal times is not always required.

The earlier onset of activity of insulin lispro is directly related to its more rapid rate of absorption into the blood from its site of injection. Unmodified insulin molecules display a strong propensity to associate into dimers as a result of interchain hydrogen bonding. In the presence of Zn three dimers then aggregate, forming a stable hexameric structure, which is characteristic of unmodified insulin pharmaceutical preparations. After SC administration, the hexamers and dimers must first dissociate before individual insulin molecules are free to enter the bloodstream. The altered amino acid sequence in the B chain of insulin lispro promotes a significant reduction in the latter's propensity to self-associate. The formulation, including Humalog in suspension with protamine sulphate, also exhibits the advantage of a rapid and a prolonged action. The formulation consisting of Humalog completely in suspension with protamine sulphate is used to control the basal level of insulin. The safety profile of insulin lispro is comparable to that of existing human insulin products.

Further Reading

http://www.eudra.org
http://www.fda.gov
http://www.humalog.com
http://www.lilly.com

Bakaysa, D. et al., Physico-chemical basis for the rapid time-action of Lys B28-ProB29-insulin: dissociation of a protein-ligand complex, *Protein Sci.*, 5, 2521–2531, 1996.

Chance, R., Glazer B., and Wishner K., Insulin L ispro (Humalog), in *Biopharmaceuticals, an Industrial Perspective,* Walsh, G. and Murphy, B., Eds., Dordrecht: Kluwer Academic Publisher, 149–171, 1999.

Ciszak, E. et al., Rate of C-terminal B-chain residues in insulin assembly: the structure of hexameric LysB28-ProB29 human insulin, *Structure,* 3, 615–622, 1995.

Galloway, J. et al., Human insulin and its modifications, in *The Clinical Pharmacology of Biotechnology Products,* Reidenberg, Ed., Amsterdam: Elsevier, 23–34, 1991.

Holleman, F. and Hoekstra, J.B., Insulin lispro, *N. Engl. J. Med.*, 337, 176–183, 1997.

Koivisto, V.A., The human insulin analogue insulin lispro, *Ann. Med.*, 30, 260–266, 1998.

HumaSPECT

Product Name:	HumaSPECT (trade name) Votumumab (international nonproprietary name)
Description:	HumaSPECT is a human monoclonal antibody against a cytokeratine tumor-associated complex of antigens that is found in colorectal adenocarcinomas. The antibody is supplied in a lyophilized form and is coupled to the radioisotope technetium 99m (^{99m}Tc) before administration as an intravenous injection.
Approval Date:	1998 (E.U.)
Therapeutic Indications:	HumaSPECT is indicated for the imaging of recurrence and metastases in patients with histologically proven colon or rectum carcinomas.
Manufacturer:	Organon Teknika, BV, Boseind 15, 5281 RM Boxtel, the Netherlands, http://www.organonteknika.com (manufacturer is responsible for import and batch release in the European Economic Area)
Marketing:	Organon Teknika, BV, Boseind 15, 5281 RM Boxtel, the Netherlands, http://www.organonteknika.com

Manufacturing

The human monoclonal antibody 88BV59, an IgG3 kappa immunoglobulin, is produced in a bioreactor using a human cell line derived from B lymphocytes. The B lymphocyte was initially sourced from a rectal adenocarcinoma and was immortalized by treatment with Epstein-Barr virus. The purification process involves several chromatographic steps, including protein G chromatography, ion exchange chromatography, as well as procedures for inactivation and removal of viruses. The final product is presented in a kit containing different components for the preparation of the solution for injection, including the lyophilized form of the antibody votumumab as well as sodium chloride, sodium hydrogen phosphate monohydrate, disodium dihydrogen phosphate heptahydrate, lactose monohydrate, stannous chloride dihydrate, D-saccharic acid monopotassium salt, sodium bicarbonate, and diethylenetriaminepentaacetic acid.

The shelf life of the product is 18 months when stored at 2 to 8°C. Routine evaluative tests (HPLC, IEF, SDS-PAGE, analysis for the presence of DNA, mycoplasma, viruses, bovine BSA, etc.) are carried out on the final product to ensure quality and safety.

Overview of Therapeutic Properties

HumaSPECT is the first completely human monoclonal antibody approved for medical use. The antibody recognizes a complex of antigens related to cytokeratines 8, 18, and 19. These are found in association with certain tumor cells, mostly colon and rectum adenocarcinomas but also breast, prostate, lung, and gastric carcinomas. After coupling of the antibody to the ^{99m}Tc, the antibody can be used for radioimaging in patients with a histologically proven colon or rectum carcinoma in order to target occult, metastatic, or recurrent tumors. HumaSPECT is administered as an intravenous injection followed by radioisotope detection within 24 hours, using a standard nuclear camera.

Results of pivotal trials show HumaSPECT to be very sensitive and accurate in diagnosis of the extent and location of colorectal carcinomas. In patients with occult carcinomas, HumaSPECT proved to be more sensitive than the currently available CT scan.

HumaSPECT was found to be safe and well tolerated. Very mild and rare side effects were observed, the most common of which were fever (experienced by 1.6% of the patients) and hypertension (occurring in 1.5% of the patients). The use of ^{99m}Tc ensures low levels of radiation to patients and a short interval between injection and imaging. The human antibody showed no immunogenicity. No human antihuman antibodies were detected, even after repeated administrations of the antibody.

HumaSPECT is contraindicated during pregnancy and lactation. No safety data are available for the use of HumaSPECT in patients under 18 years old. Ongoing clinical trials are monitoring the use of HumaSPECT in targeting ovarian and prostate tumors.

Further Reading

http://www.eudra.org
http://www.organonteknika.com

Serafini, A.N. et al., Radioimmunoscintigraphy of recurrent, metastatic, or occult colorectal cancer with technetium 99m-labeled totally human monoclonal antibody 88BV59: results of pivotal, phase III multicenter studies, *J. Clin. Oncol.*, 16, 1777–1787, 1998.

Wolff, B. G. et al., Radioimmunoscintigraphy of recurrent, metastatic, or occult colorectal cancer with Technetium Tc 99M 88BV 59H21 2V67 66 (Humaspect), a totally human monoclonal antibody. Patient management benefit from a phase III multicenter study, *Dis. Colon Rectum*, 41(8), 953–962, 1998.

Humatrope

Product Name: Humatrope (trade name)

Somatropin (international nonproprietary name)

Description: Humatrope is a recombinant human growth hormone (hGH, somatropin). The 22.1-kDa, 191-amino acid single-chain polypeptide is produced in a modified *Escherichia coli* strain using recombinant DNA technology. Humatrope is supplied in lyophilized form (5-mg product, equivalent to 15-IU activity). A solvent for reconstitution (in vials, or in cartridges for use with the multidose device HumatroPen) is also provided. The product is administered by subcutaneous or intramuscular injection.

Approval Date: 1987 (U.S.)

Therapeutic Indications: Humatrope is indicated for the treatment of children with a growth failure due to insufficient endogenous growth hormone or due to Turner's syndrome. It is also now indicated for the treatment of adults with a growth hormone deficiency as a result of pituitary or hypothalamic disease, surgery, radiotherapy, trauma, or a growth hormone deficiency during childhood. Those patients should have previously failed to respond to a standard growth hormone stimulation test.

Manufacturer: Eli Lilly and Co., Lilly Corporate Center, Indianapolis, IN 46285, http://www.lilly.com

Marketing: Eli Lilly and Co., Lilly Corporate Center, Indianapolis, IN 46285, http://www.lilly.com

Manufacturing

Humatrope is produced by recombinant DNA technology in *E. coli* cells. It is initially synthesized with an additional methionine residue at the N-terminus, but this is cleaved as the polypeptide is transported naturally into the cell's periplasmic space. Biologically active somatropin, identical to the native human molecule, is extracted from the periplasmic space subsequent to fermentation. The molecule is then purified using a number of chromatographic steps. The final product is presented in lyophilized form in vials and cartridges, with a solvent to be used in reconstitution prior to subcutaneous or intramuscular administration. The cartridges are used with a multidose device called HumatroPen.

Humatrope consists of somatropin (active), as well as mannitol, glycine, dibasic sodium phosphate, phosphoric acid, and sodium hydroxide as excipients. The solvent consists of metacresol and glycerin in water for injection. Tests are carried out to ensure the quality and safety of the product.

Overview of Therapeutic Properties

Humatrope is a recombinant growth hormone that is used in growth hormone replacement therapy. It is indicated for treatment of children with growth failure due to lack of adequate levels of the endogenous hormone and for treatment of short stature associated with Turner's syndrome in patients whose epiphyses are not closed. It is also indicated for the treatment of adults who have not responded to a standard growth hormone stimulation test and who exhibit a growth hormone deficiency that results from pituitary or hypothalamic disease, surgery, radiotherapy, trauma, or growth hormone deficiencies during childhood.

Humatrope exhibits the same activity as the human growth hormone of pituitary origin. It should be administered subcutaneously or intramuscularly at doses adjusted to individual patient needs, daily or on 3 alternate days. Clinical trials showed that children who received Humatrope long-term showed an increase in height of about 5 cm compared with children who received placebos. Patients with Turner's syndrome achieved height increases of 5 to 8 cm. Humatrope administration in adults led to an increase in lean body mass, a decrease in fat, and an increase in exercise capacity and work performance.

Pain at the injection site and mild and transient edema were the most commonly reported, though infrequent, side effects. Antibodies against somatropin were also reported. A small number of cases of leukemia have been observed, but a relationship with Humatrope was not proven. Humatrope should not be administered during pregnancy and lactation or in patients with malignancies or acute critical illness due to complications after open heart surgery, abdomen surgery, or multiple accidental trauma.

Further Reading

http://www.fda.gov
http://www.humatrope.com
http://www.lilly.com
Alemzadeh, R. et al., Anabolic effects of growth hormone treatment in young children with cystic fibrosis, *J. Am. Coll. Nutr.*, 17(5), 419–424, 1998.
Daniels, M.E., Lilly's Humatrope experience, *Bio-Technol.*, 10(7), 812, 1992.
Fernholm, G. et al., Growth hormone replacement therapy improves body composition and increases bone metabolism in eldery patients with pituitary disease, *J. Clin. Endocr. Metab.*, 85(11), 4104–4112, 2000.
Kawashige, M. et al., Quality evaluation of commercial lyophilized human growth hormone preparations, *Biol. Pharm. Bull.*, 18(12), 1793–1796, 1995.

Humira

Product Name: Humira (trade name)

Adalimumab (common name)

Description: Humira contains as an active ingredient a recombinant human IgG1 monoclonal antibody that specifically binds to human tumor necrosis factor alpha (TNF-α). It does not bind, or in any way affect, TNF-β (lymphotoxin). Humira is the first monoclonal antibody-based product approved to be created using phage display technology. The 148-kDa, 1330-amino acid antibody is produced by recombinant DNA technology and is supplied in a single-use, 1-ml glass syringe for subcutaneous use. Each syringe delivers 0.8 ml product, which contains 40 mg active substance.

Approval Date: 2002 (U.S.); 2003 (E.U.)

Therapeutic Indications: Humira is indicated for reducing signs and symptoms and inhibiting the progression of structural damage in adult patients with moderately to severely active rheumatoid arthritis who have had an inadequate response to one or more disease-modifying antirheumatic drugs (DMARDs). It can be used alone or in combination with methotrexate or other DMARDs.

Manufacturer: Abbott Laboratories, 1401 Sheridan Road, North Chicago, IL 60064-4000, http://www.abbott.com (U.S.)
Abbott GmbH & Co., Max Planck — Ring 2, D-65205, Wiesbaden, Germany (manufacturer is responsible for batch release in the European Economic Area)

Marketing: Abbott Laboratories, 1401 Sheridan Road, North Chicago, IL 60064-4000 (U.S.)
Abbott Laboratories Ltd., Queenborough, Kent, ME 5EL U.K. (E.U.)

Manufacturing

Humira contains as active ingredient an IgG1 human antibody created using phage display technology. It is produced by recombinant DNA technology in an engineered mammalian cell line. Following cell culture, the extracellular product is recovered and subject to a number of high-resolution chromatographic purification steps. Also included in the downstream processing procedures are specific steps to inactivate and remove any viruses potentially present in the product stream. The final product is formulated as a sterile, preservative-free solution for subcutaneous administration.

In addition to the active ingredient, Humira contains sodium chloride, mannitol, polysorbate 80, sodium phosphate, sodium citrate, and citric acid as excipients. Sodium hydroxide is added as necessary to adjust the final product pH to 5.2. The product is filter-sterilized and filled into single-use, 1-ml glass syringes. It is stored between 2 and 8°C, protected from light.

Overview of Therapeutic Properties

Rheumatoid arthritis (RA) is an autoimmune condition in which joints become inflamed and painful. The condition develops in approximately 1% of the population, mostly in women 25 to 50 years old. There are approximately 2 million sufferers in the U.S. alone. As the immune condition progresses, cartilage, bone, and ligaments of the joints erode, which can eventually severely disable 10% of sufferers. Severe RA is often treated with immunosuppressive drugs, such as methotrexate, azathioprine, and cyclophosphamide.

A major feature of RA is the presence of elevated levels of the proinflammatory cytokine TNF-α in the joints. TNF-α plays a central role in fueling pathologic inflammation and joint destruction. Humira reduces the signs and symptoms of RA and inhibits the progression of structural damage by binding to TNF-α. Binding prevents initiation of TNF-αs biological effects by preventing it from binding to the P55 and P75 TNF cell surface receptors. After its administration, a rapid decrease in various serum cytokines — as well as acute phase reactants of inflammation (C-reactive protein) — is observed.

The product, which is usually self-administered once every other week, reduces the signs and symptoms of RA in more than 50% of patients. The product can elicit a number of potentially serious side effects, including increased incidence of serious, sometimes life-threatening infections, including tuberculosis. Patients should be evaluated for latent tuberculosis infection with a tuberculin skin test prior to commencement of therapy. Additional rare side effects include nervous system-related symptoms (e.g., numbness, dizziness, problems with vision) as well as a potentially increased risk of developing various malignancies and a lupus-like syndrome.

Further Reading

http://www.abbottimmunology.com
http://www.fda.gov
http://www.humira.com
Anon., Adalimumab (Humira) for rheumatoid arthritis, *Med. Lett. Drugs Ther.*, 45(1153), 25–27, 2003.
Anon., Humira — Adalimumab — Abbott laboratories — new treatment for rheumatoid arthritis, *Formulary*, 38(2), 78–79, 2003.
Braun, J. and Van der Heijde, D., Novel approaches in the treatment of ankylosing spondylitis and other spondyloarthritides, *Expert Opin. Inv. Drug.*, 12(7), 1097–1109, 2003.
Edwards, L.A., Adalimumab, a fully human monoclonal antitumor necrosis factor alpha antibody, *Formulary*, 38(5), 272–277, 2003.
Souriau, C. and Hudson, P.J., Recombinant antibodies for cancer diagnosis and therapy, *Expert Opin. Biol. Th.*, 3(2), 305–318, 2003.

Humulin

Product Name:	Humulin (trade name) Human insulin (rDNA origin) (international nonproprietary name)
Description:	Humulin is a recombinant human insulin produced in *Escherichia coli* using recombinant DNA technology. It is provided in different formulations with different onsets and durations of activity: Humulin R (regular), Humulin N (NPH; isophane), Humulin L (Lente), Humulin U (UltraLente), Humulin 50/50, and Humulin 70/30. Humulin N and Humulin 70/30 are available with the multidose Humulin Pen.
Approval Date:	1982 (U.S.)
Therapeutic Indications:	Humulin is used to control the blood glucose level of patients with diabetes mellitus.
Manufacturer:	Eli Lilly and Company, Lilly Corporate Center, Indianapolis, IN 46285, http://www.lilly.com
Marketing:	Eli Lilly and Company, Lilly Corporate Center, Indianapolis, IN 46285, http://www.lilly.com

Manufacturing

Humulin is a recombinant human insulin produced using recombinant DNA technology. Initially, Lilly produced Humulin using a manufacturing method developed by Genentech, which consisted of separate production and purification of the insulin A and B chains. The purified chains were subsequently covalently linked to produce a native insulin molecule. Since 1986, a new manufacturing method has been used in which proinsulin (the natural insulin precursor, which contains both A and B chains linked by a C or "connecting" peptide) is expressed in *E. coli*. The cloning strategy undertaken results in the expression of a chimeric molecule, consisting of the enzyme tryptophan synthesase linked to proinsulin via a methionine residue. The proinsulin is released (via cyanogen bromide-mediated hydrolysis) and purified. Mature insulin is then released via treatment with carboxypeptidase B and trypsin.

The insulin is chromatographically purified and packaged in a number of different formulations: Humulin is supplied as Humulin R (regular), with zinc insulin crystals dissolved in solution; Humulin N (NPH; isophane) is a suspension of zinc insulin with protamine sulphate; Humulin L (Lente) is an amorphous and crystalline zinc suspension; Humulin U (UltraLente) is a crystalline suspension of zinc insulin; Humulin 50/50 is a half-and-half mixture of insulin and insulin isophane; and Humulin 70/30 is a mixture of 70% insulin isophane and 30% insulin regular. Humulin N and Humulin 70/30 are available with the multidose Humulin Pen.

Overview of Therapeutic Properties

Humulin exhibits the same activity as the naturally occurring human insulin and is used to promote glucose homeostasis in patients with diabetes mellitus. It is offered in several formulations, which differ in their onset of action and duration of activity. Humulin R (regular) consists of zinc insulin crystals dissolved in a clear solution and exhibits a relatively short duration of activity (4 to 12 hours), with a relatively rapid onset of action that peaks 2 to 5 hours after administration. Humulin N (NPH; isophane), a suspension of zinc insulin with protamine sulphate, is an intermediate-acting insulin with a slower onset of action and a longer duration of activity (up to 24 hours). Humulin L (Lente), a suspension of amorphous and crystalline zinc insulin, has a slower onset of activity but a duration of action of up to 24 hours. Humulin U (UltraLente), a crystalline suspension of zinc insulin, has a very slow onset of action and a duration of activity of up to 28 hours. Humulin 70/30 and Humulin 50/50 are mixtures of 70% and 50%, respectively, of insulin isophane mixed with regular insulin, and combine the rapid onset of action of regular insulin with the duration of action of up to 24 hours of insulin isophane.

Humulin is usually administered subcutaneously at a dosage adjusted to the individual needs of the patient. Humulin is well tolerated and exhibits very mild side effects. Hypoglycemia, reactions at the injection site, and lipodystrophy have been reported.

Further Reading

http://www.fda.gov
http://www.humulin.com

http://www.lilly.com
Anon., Bye bye humulin, *Diabetes Obes. Metab.*, 3(1), 59, 2001.
Aristides, M. et al., Patient preference and willingness to pay in 5 European countries for humalog mix 25 compared to humulin 30/70 for the treatment of type 2 diabetes, *Value Health*, 5(6), 448, 2002.
Hollerman, F. et al., Pharmacokinetics of insulin lispro mid mixture compared to humulin 50/50, *Diabetologi*, 41, 937, 1998.
Takenaga, M. et al., A novel insulin formulation can keep providing steady levels of insulin for much longer periods *in vivo*, *J. Pharm. Pharmacol.*, 54(9), 1189–1194, 2002.

Indimacis 125 (withdrawn from market)

Product Name: Indimacis 125 (trade name)

Igovomab (international nonproprietary name)

Description: The active substance of indimacis 125 (Igovomab) consists of an antigen-binding $[F(ab)_2]$ fragment of a murine IgG monoclonal antibody raised against the CA 125 antigen. The antibody fragment is covalently linked to the chelating agent diethylenetriamine penta-acetic acid (DTPA). The product is provided as a solution in an ampule and contains 1 mg/ml Igovomab. Radioactive indium must be purchased separately and chelated to Igovomab immediately prior to use.

Approval Date: 1996 (E.U.)

Withdrawal Date: This product is no longer on the market. Although not listed on the European Medicines Evaluation Agency's (EMEA's) list of product withdrawals, its European public assessment report is no longer available on the EMEA Web site, and it is not listed on the European Commission's register of approved medicinal products.

Therapeutic Indications: Indimacis 125 was indicated for positive diagnosis of relapsing ovarian adenocarcinoma when serum CA 125 is increased without positive results of ultrasound or computerized tomography scan.

Manufacturer: Cis bio International, B.P. 32-91192, Gif sur Yvette, Cedex, France (manufacturer was responsible for batch release in the European Economic Area)

Marketing: Cis bio International, B.P. 32-91192, Gif sur Yvette, Cedex, France

Manufacturing

The monoclonal antibody is produced in a mouse hybridoma cell line. After cell culture, the intact murine antibody is proteolytically cleaved with pepsin, yielding the antigen-binding Fab_2 antibody fragment. This is further chromatographically purified and covalently linked to DTPA. After excipient addition the product is filter sterilized and aseptically filled into 1-ml single-dose glass ampules.

Overview of Therapeutic Properties

CA 125 is an oncofoetal protein expressed by more than 90% of ovarian serous adenocarcinomas. Diagnosis of relapse of ovarian cancer often relies upon ultrasound imaging or computerized tomography, which can lack diagnostic sensitivity. Administration of Indimacis 125 labeled with radioactive indium (bound to the antibody via the DTPA group) provides a sensitive immunoscintigraphic imaging method for detecting relapses. Trials illustrate that, compared to ultrasound or computerized tomography scans, Indimacis 125 scintigraphy is a positive determinant in about 50% of the cases where the other imaging techniques were negative. No severe side effects relating to allergic reactions were noted.

InductOs

Product Name: InductOs (trade name)

Dibotermin alfa (international nonproprietary name)

Description: InductOs consists of a recombinant human bone morphogenetic protein-2 (h BMP-2), which promotes the differentiation of mesenchymal cells into bone cells. BMP-2 is a member of the transforming growth factor beta (TGF-β) superfamily of growth and differentiation factors. The biologically active form is a glycosylated heterodimer, consisting of 114- and 131-amino acid polypeptide subunits. It is produced in a Chinese hamster ovary (CHO) cell line using recombinant DNA techniques. The product is presented as a lyophilized powder (12 mg/vial) with a solvent for reconstitution and a matrix made of bovine collagen for keeping the active substance in place following peri-osseous application during surgical procedures.

Approval Date: 2002 (E.U.)

Therapeutic Indications: InductOs is indicated in skeletally mature patients for the treatment of acute tibia fractures in adjunct to standard care using fracture reduction and intramedullary nail fixation.

Manufacturer: Wyeth Laboratories, New Lane, Havant, Hants PO9 2NG, U.K., http://www.wyeth.com (manufacturer is responsible for import and batch release in the European Economic Area)

Marketing: Genetics Institute of Europe B.V., Fraunhoferstrasse 15, 82152 Planegg/Martinsried, Germany, http://www.wyeth.com

Manufacturing

InductOs contains the active substance dibotermin alfa, which is the recombinant h BMP-2. Dibotermin alfa is produced by recombinant DNA technology in CHO cells. The cloned gene was amplified from human osteosarcoma cells using a bovine genomic BMP-2 fragment. The glycosylated dimeric protein is excreted in the culture medium and purified using several chromatographic and filtration procedures. InductOs is presented as a kit including the active substance dibotermin alfa in a lyophilized form, a solvent (water for injection) for reconstitution, and a matrix, an absorbable type-1 collagen sponge of bovine origin. Sucrose, glycine, glutamic acid, sodium chloride, sodium hydroxide, and polysorbate 80 are included as excipients.

The shelf life of the product is 24 months when stored between 15 and 30°C. Tests carried out to ensure the quality and the safety of the product include extensive characterization of the protein and viral and microbial analysis.

Overview of Therapeutic Properties

InductOs is indicated in the treatment of fractures of the tibia in adult patients. It is used in combination with standard care using fracture reduction and intramedullary nail fixation. The active substance of InductOs, dibotermin alfa, is a recombinant h BMP-2, which binds to receptors on the surface of mesenchymal cells and induces differentiation into bone cells. InductOs is applied during surgical procedures at the site of fracture. The use of a bovine collagen sponge ensures retention of the active substance at the site of the fracture for the time required for healing, with the matrix dissolving over time until undetectable. A clinical trial demonstrated the efficacy of InductOs in inducing a more rapid healing of fractures and a reduced need for additional surgery to heal fractures, compared to standard treatment. No difference in the rate of healing was observed in patients who received a reamed intramedullary nail fixation.

The most commonly reported side effects related to the use of InductOs were headache, pancreatic malfunction, transient tachycardia, and a low level of magnesium in the blood, while a reduction in pain and infections at the fracture site was observed. Antibodies against dibotermin were observed but did not affect the efficacy of the treatment. InductOs is contraindicated during pregnancy and in the presence of malignancies.

Further Reading

http://www.eudra.org
http://www.wyeth.com
Govender, S. et al., Recombinant human bone morphogenetic protein-2 for the treatment of open tibial fractures — a prospective controlled, random study of 450 patients, *J. Bone Joint Surg. Am.*, 84A(12), 2123–2134, 2002.
Riedel, G.E. and Valentin-Opran, A., Clinical evaluation of rhBMP-2/ACS in orthopedic trauma: a progress report, *Orthopedics*, 22, 663–665, 1999.

INFANRIX HepB

Product Name: INFANRIX HepB (trade name)

Diphteria, tetanus, acellular pertussis, and hepatitis B vaccine (international nonproprietary name)

Description: INFANRIX HepB is a vaccine that combines an already existing trivalent vaccine against diphtheria, tetanus, and acellular pertussis (trade name INFANRIX) with a hepatitis B vaccine produced in yeast by recombinant DNA technology. INFANRIX HepB consists of diphtheria and tetanus toxoids, three pertussis antigens (pertussis toxoid, filamentous hemagglutinin, and pertactin), and recombinant major surface antigen of the hepatitis B virus. INFANRIX HepB is supplied as a suspension of the diphtheria, tetanus, and pertussis components, which is used to reconstitute the lyophilized hepatitis B component before administration as an intramuscular injection.

Approval Date: 1997 (E.U.)

Therapeutic Indications: INFANRIX HepB is indicated for active immunization against diphtheria, tetanus, pertussis, and hepatitis B in infants from the age of 2 months.

Manufacturer: GlaxoSmithKline, Rue de l'Institut 89, 1330 Rixensart, Belgium, http://www.gsk.com (manufacturer is responsible for import and batch release in the European Economic Area)

Marketing: GlaxoSmithKline, Rue de l'Institut 89, 1330 Rixensart, Belgium, http://www.gsk.com

Manufacturing

INFANRIX HepB is a combination of two vaccines already approved in the E.U. — INFANRIX (a trivalent vaccine against diphtheria, tetanus, and pertussis)

and Engerix-B (a recombinant vaccine against hepatitis B). The components are obtained and purified separately and are subsequently combined. The diphtheria toxoid is obtained from a culture of *Corynebacterium diphtheriae* and the tetanus toxin from a culture of *Clostridium tetani*. After a detoxification treatment with formaldehyde, the toxins are purified by salt fractionation, ultradialysis, and filtrations. The pertussis toxin, filamentous hemagglutinin, and pertactin, antigens used to elicit immunogenicity against pertussis, are obtained from a culture of *Bordetella pertussis*. After extraction and purification, the antigens are treated with formaldehyde (pertussis toxin is also treated with glutaraldehyde). The surface antigen of the hepatitis B virus is produced by recombinant DNA technology in yeast cells (*Saccharomyces cerevisiae*) and purified in the form of nonglycosylated polypeptides, which retain the characteristics of the natural antigen. Diphtheria, tetanus, and pertussis components are adsorbed to aluminium hydroxide and presented as a suspension, whereas the hepatitis component, adsorbed to aluminium phosphate, is presented separately in a lyophilized form to be reconstituted with the suspension before administration as an intramuscular injection. INFANRIX HepB contains the active immunogenic components, as well as aluminium hydroxide, aluminium phosphate, formaldehyde, 2-phenoxyethanol (an antimicrobial agent), polysorbate 20 and 80, and sodium chloride (for isotonicity).

The shelf life of the product is 36 months when stored at 2 to 8°C, protected from light. Tests carried out to ensure the quality and safety of the product include ELISA, SDS-PAGE, and RIA.

Overview of Therapeutic Properties

INFANRIX HepB is indicated for the immunization of infants against diphtheria, tetanus, pertussis, and hepatitis B. It consists of purified antigenic components, which elicit the immunoprotection, with the advantage (over the whole-cell vaccines) of reduced immunogenicity. The combination of more vaccines into one preparation facilitates administration.

INFANRIX HepB should be administered as a three-dose vaccine with the first dose at 2 months of age and the other doses at 2-month intervals, or with the first dose administered at 3 months of age and the subsequent doses at intervals of at least 1 month. A booster dose could be administered before the end of the second year. INFANRIX HepB can be administered simultaneously with polio and *Haemophilus influenzae* vaccines.

INFANRIX HepB proved to elicit an immunological defense to the same extent as in separate and simultaneous administrations of triavalent vaccine against diphtheria, tetanus, pertussis, and the hepatitis B vaccine. INFANRIX HepB was found to be safe and well tolerated, with the same side effects observed as after administration of the separate vaccines. Caution should be exercised in the case of infants with thrombocytopenia or a bleeding disorder. It should not be administered in the case of fever.

Further Reading

http://www.eudra.org

http://www.gsk.com

Yea, S. et al., Safety and immunogecity of a pentavalent diphtheria, tetanus, pertussis, hepatitis B and polio combination vaccine in infants, *Pediatr. Infect. Dis. J.*, 20(10), 973–980, 2001.

Infanrix Hexa

Product Name: Infanrix Hexa (trade name)

Diphtheria, tetanus, acellular pertussis, hepatitis B recombinant (adsorbed), inactivated poliomyelitis, adsorbed conjugated ***Haemophilus influenzae*** type b (Hib) vaccine (international nonproprietary name)

Description: Infanrix Hexa is a combined vaccine against diphtheria, tetanus, pertussis, hepatitis B, poliomyelitis and Hib. It is a combination of vaccines already approved in the E.U. Infanrix Hexa consists of diphtheria and tetanus toxoids, three pertussis antigens (pertussis toxoid, filamentous hemagglutinin, and pertactin), the major surface antigen of the hepatitis B virus obtained by recombinant DNA technology, inactivated polioviruses (IPV) (types 1, 2, and 3), and Hib polysaccharide. Infanrix Hexa is presented as a suspension, containing diphtheria, tetanus, pertussis, hepatitis B, and poliomyelitis components, which is used to resuspend the lyophilized *H. influenzae* component before administration as an intramuscular injection.

Approval Date: 2000 (E.U.)

Therapeutic Indications: Infanrix Hexa is indicated for primary and booster vaccination against diphtheria, tetanus, pertussis, hepatitis B, poliomyelitis, and Hib in infants.

Manufacturer: GlaxoSmithKline, Rue de l'institut 89, 1330 Rixensart, Belgium, http://www.gsk.com (manufacturer is responsible for import and batch release in the European Economic Area)

Marketing: GlaxoSmithKline, Rue de l'institut 89, 1330 Rixensart, Belgium, http://www. gsk.com

Manufacturing

Infanrix Hexa consists of different immunogenic components produced and purified separately and subsequently combined. Diphtheria and tetanus toxoids are obtained from cultures of *Corynebacterium diphtheria* and *Clostridium tetani*, respectively. After treatment with formaldehyde, both toxins are purified using salt fractionation, dialysis, and filtrations. Pertussis toxin, filamentous hemagglutinin, and pertactin are treated with formaldehyde (pertussis toxin is also treated with glutaraldehyde) after extraction and purification from a culture of *Bordetella pertussis*. The surface antigen of the hepatitis B virus is produced by recombinant DNA technology in *Saccharomyces cerevisiae*. IPV is produced by culture in Vero cells and, after purification, is inactivated with formaldehyde. Hib polysaccharide is extracted from the *H. influenzae* strain 20,752, activated with cyanogen bromide and adipic acid hydrazide, then coupled to tetanus toxoid, adsorbed on aluminum salt to increase immunoreactivity, and lyophilized.

Diphtheria, tetanus, and pertussis components are adsorbed on hydrated aluminum oxide, hepatitis B, and Hib components on aluminum phosphate. Inactivated polioviruses are adsorbed on aluminum after mixing with the other components. All components other than Hib are combined in a suspension that is used to resuspend the lyophilized Hib component before administration. Infanrix Hexa contains the antigenic components, anhydrous lactose (as a stabilizer for the Hib component), sodium chloride (as an isotonic agent), phenoxyethanol (as a preservative), hydrated aluminum oxide and aluminium phosphate (as adjuvants), medium 199 containing mainly amino acids, mineral salts, and vitamins (as a stabilizer for the IPV component), and traces of the antibiotics neomycin and polymyxin.

The shelf life of the product is 24 months when stored at 2 to 8°C, protected from light. Tests to ensure the quality and safety of the vaccine include assays to detect viruses and toxins, and potency and *in vitro* antigenic assays.

Overview of Therapeutic Properties

The combined vaccine Infanrix Hexa is indicated for primary and booster vaccination of infants against diphtheria, tetanus, pertussis, hepatitis B, poliomyelitis viruses, and Hib. It should be administered as a three-dose vaccine, with at least a 1-month interval between the injections.

Infanrix Hexa is a combination of already approved vaccines, and studies on primary vaccinations showed that it is as safe and well tolerated as when the compounds are administered separately. Infanrix Hexa results in the same immunogenicity as vaccines given separately, except for the Hib component. The titre of anti-Hib antibodies is lower than the titre observed if the Hib vaccine is given separately. This well-known phenomenon occurs with combined vaccines and is due to an interference with the pertussis vaccine. However, the immunodefense against Hib is not affected. Studies proved that, even in the presence of a low titre of anti-Hib antibodies, the immunodefense is the same as for the separated vaccine and still exists after at least 6 years from vaccination. This suggests that immunomemory, more than a humoral immunoresponse, is responsible for long-term protection.

An increased reactogenicity was observed after booster vaccination with diphtheria, tetanus, and pertussis or combinations thereof, resulting in a higher incidence of fever than in primary vaccination. Caution should be exercised in infants with thrombocytopenia or a bleeding disorder. It should not be administered in the case of fever.

Further Reading

http://www.eudra.org

http://www.gsk.com

Gylca, R. et al., A new DTPa-HBV-IPV vaccine co-administrered with Hib, compared to a commercially available DTPw-IPV/Hib vaccine co-administrered with HBV, given at 6, 10, and 14 weeks following HBV at birth, *Vaccine*, 19, 825–833, 2000.

Schmitt, H.J. et al., Primary vaccination of infants with diphtheria, tetanus, acellular pertussis, hepatitis B virus, inactivated polio virus, and Haemophilus influenzae type b vaccines given as either separate or mixed injections, *J. Pediatr.*, 137, 304–312, 2000.

Infanrix Penta/Pediarix

Product Name: Infanrix Penta (trade name E.U.)

Pediarix (trade name U.S.)

Diphtheria, tetanus, acellular pertussis, hepatitis B recombinant (adsorbed), and inactivated poliomyelitis vaccine (international nonproprietary name)

Description: Infanrix Penta (Pediarix in the U.S.) is a combined vaccine against diphtheria, tetanus, pertussis, hepatitis B, and poliomyelitis. It is a combination of vaccines already approved (Infanrix and Engerix-B in the U.S., Infanrix HepB, and Infanrix IPV in the E.U.). Infanrix Penta contains diphtheria and tetanus toxoids, three pertussis antigens (pertussis toxoid, filamentous hemagglutinin, and pertactin), the major surface antigen of the hepatitis B virus obtained by recombinant DNA technology, and inactivated polioviruses (IPV) (types 1, 2, and 3). It is presented as a suspension in prefilled syringes for intramuscular injection.

Approval Date: 2000 (E.U.); 2002 (U.S.)

Therapeutic Indications: This product is indicated for primary and booster vaccinations against diphtheria, tetanus, pertussis, hepatitis B, and poliomyelitis in infants.

Manufacturer: GlaxoSmithKline Biologicals S.A., Rue de l'Institut 89, 1330 Rixensart, Belgium, http://www.gsk.com (E.U. and U.S.)

Marketing: GlaxoSmithKline Biologicals S.A., Rue de l'Institut 89, 1330 Rixensart, Belgium, http://www.gsk.com (E.U.)

GlaxoSmithKline Pharmaceuticals, One Franklin Plaza, Philadelphia, PA 19102, http://www.gsk.com (U.S.)

Manufacturing

Infanrix Penta consists of a variety different immunogenic components. Diphtheria and tetanus toxoids are obtained from *Corynebacterium diphtheria* and *Clostridium tetani*, respectively. Both toxins are inactivated using formaldeyde and are subsequently purified. Pertussis toxin, filamentous hemagglutinin, and pertactin, derived from *Bordetella pertussis*, are purified and inactivated by formaldehyde treatment (pertussis toxin is also treated with glutaraldehyde). The surface antigen of the hepatitis B virus is produced by recombinant DNA technology in *Saccharomyces cerevisiae*. IPV is obtained from a culture of Vero cells, a continuous line of monkey kidney cells, and, after purification, is inactivated with formaldehyde. Diphtheria, tetanus, and pertussis components are adsorbed on hydrated aluminium oxide, while the hepatitis B component is adsorbed on aluminium phosphate. Inactivated polioviruses are adsorbed on aluminium after mixing with the other components. Infanrix Penta is presented as a suspension for intramuscular injection and contains the antigenic components, sodium chloride (as an isotonic agent), hydrated aluminium oxide and aluminium phosphate (as adjuvants), phenoxyethanol (as a preservative), and medium 199 containing mainly amino acids, mineral salts, and vitamins (as a stabilizer for the IPV component).

The shelf life of the product is 36 months when stored at 2 to 8°C, protected from light. Extensive testing carried out to ensure the quality and safety of the vaccine include toxicity and potency tests.

Overview of Therapeutic Properties

Infanrix Penta is indicated for primary and booster vaccination of infants against diphtheria, tetanus, pertussis, hepatitis B, and the poliomyelitis viruses. Infanrix Penta is a combination of previously approved vaccines and was shown to be as safe and well tolerated as the separately administered vaccines. Reactions at the injection site and flu-like symptoms were the most commonly reported side effects. A higher rate of increased fever was reported after administration of Infanrix Penta than after administration of the separate vaccines.

Infanrix Penta proved to be effective in eliciting an immunoresponse against diphtheria, tetanus, pertussis, hepatitis B, and polioviruses. Studies showed that Infanrix Penta could be administered simultaneously with vaccines against *Haemophilus influenzae* type b. Infanrix Penta should be administered as a three-dose vaccine with a 1- or 2-month interval between injections, from the age of 2 months. It should not be administered to infants under 6 weeks old or in the case of fever.

Further Reading

http://www.eudra.org
http://www.fda.com
http://www.gsk.com
http://www.pediarix.com
Yeh, S.H. et al., Safety and immunogenicity of a pentavalent diphtheria, tetanus, pertussis, hepatitis B, and polio combination vaccine in infants, *Pediatr. Infect. Dis. J.*, 20, 973–980, 2001.

Infergen

Product Name: Infergen (trade name)

Interferon alfacon-1 (international nonproprietary name)

Description: Infergen is a 19.5-kDa, 166-amino acid single-chain non-glycosylated polypeptide produced in recombinant form in *Escherichia coli*. It is a synthetic type-1 interferon. The synthetic gene consists of a consensus sequence, incorporating the most frequently observed amino acid residues at any given position in several type-1 interferon alfa sequences. Infergen is presented as a solution (9 µg in 0.3 ml) to be administered subcutaneously.

Approval Date: 1997 (U.S.); 1999 (E.U.)

Therapeutic Indications: Infergen is indicated in the treatment of adult patients with chronic hepatitis C.

Manufacturer: CFP Companhia Farmaceutica S.A., R. Consiglieri Pedroso 123, Queluz de Baixo, 2745 Barcarena, Portugal (manufacturer is responsible for batch release in the European Economic Area)
Amgen, Inc., Thousand Oaks, CA, http://www.amgen.com (U.S.)

Marketing: Yamanouchi Europe B.V., Elisabethhof 19, 2353 EW Leiderdorp, the Netherlands, http://www.yamanouchi.com (E.U.)
Amgen, Inc., One Amgen Drive Center, Thousand Oaks, CA 91320-1799, http://www. amgen.com (U.S.)

Manufacturing

Infergen is a non-natural interferon produced by recombinant DNA technology in *E. coli*. The expressed protein exhibits activity comparable to other alfa interferons, even in its unglycosylated form. After fermentation, the *E. coli* cells are collected and homogenized. Solubilization, oxidation, and refolding steps ensue, generating the crude, biologically active interferon.

Purification entails the use of several chromatographic steps, as well as procedures to remove viral and microbial contaminants. Infergen consists of the active substance interferon alfacon-1 and sodium chloride, monobasic sodium phosphate, and dibasic sodium phosphate as excipients. The final product is presented as a solution for subcutaneous injection.

The shelf life of the product is 24 months when stored at 2 to 8°C. Extensive tests carried out to ensure the quality and the safety of the product include amino acid sequence analysis, peptide mapping, mass spectrometry, isoelectric focusing, HPLC, reverse-HPLC, SDS-PAGE, Western blotting, and assays for the presence of viral and microbial contaminants.

Overview of Therapeutic Properties

Infergen is indicated in the treatment of adult patients with chronic hepatitis C infections who exhibit serum markers for the virus. The synthetic molecule retains the antiviral, antiproliferation, and immunomodulatory activity of natural interferons. Interferon alfacon-1 led to an increased antiviral activity *in vitro* but exhibited an activity comparable to natural interferon molecules in clinical trials.

Infergen should be administered subcutaneously three times a week for a period of 12 months. Infergen was found to be as effective as interferon alfa -2b in the treatment of chronic hepatitis. The side effects reported after use of Infergen are similar to those observed after administration of other interferons, including flu-like symptoms. Cardiovascular and psychiatric disorders have occasionally been reported, as well as the development of neutralizing anti-interferon antibodies in 4% of patients. Infergen is contraindicated during pregnancy and lactation and in the case of severe psychiatric, cardiovascular, renal, and hepatic disorders.

Further Reading

http://www.amgen.com
http://www.eudra.org
http://www.fda.gov
http://www.infergen.com
http://www.yamanouchi.com

Blatt, L.M. et al., The biologic activity and molecular characterization of a novel synthetic interferon-alpha species, consensus interferon, *J. Interferon Cytokine Res.*, 16, 489–499, 1996.

Melian, E.B. and Plosker, G.L., Interferon alfacon-1: a review of its pharmacology and therapeutic efficacy in the treatment of chronic hepatitis C, *Drugs*, 61, 1661–1691, 2001.

InFUSE Bone Graft/LT-CAGE Lumbar Tapered Fusion Device

Product Name: Device (trade name)

InFUSE Bone Graft/LT-CAGE Lumbar Tapered Fusion

Bone morphogenic protein-2 (common name dibotermin alfa), housed in an absorbable collagen sponge scaffold with metal prosthesis

Description: InFUSE Bone Graft/LT-CAGE Lumbar Tapered Fusion Device consists of the recombinant human bone morphogenic protein-2 (rhBMP-2), which stimulates new bone formation, combined with a bovine type I collagen sponge matrix serving as a scaffold. The bone graft component is contained in a metallic tapered fusion cage. The rhBMP-2 is produced in Chinese hamster ovary (CHO) cells using recombinant DNA technology. BMP-2 is a disulphide-linked heterodimeric protein. Both the 114- and the 131-amino acid polypeptide subunits are glycosylated. The protein is provided as a lyophilized powder to be reconstituted (to a strength of 1.5 mg/ml) before application on the provided absorbable collagen sponge. The metallic cage, available in different sizes, is provided separately.

Approval Date: 2002 (U.S.)

Therapeutic Indications: InFUSE Bone Graft/LT-CAGE Lumbar Tapered Fusion Device is indicated for spinal fusion procedures in skeletally mature patients suffering degenerative disc disease in the lower region of the spine (L4-S1).

Manufacturer: Medtronic Sofamor Danek, 1800 Pyramid Place, Memphis, TN 38132, http://www.medtronicsofamordanek.com (U.S.)

Marketing: Medtronic Sofamor Danek, 1800 Pyramid Place, Memphis, TN 38132, http://www.medtronicsofamordanek.com (U.S.)

Manufacturing

The rhBMP-2 is produced in CHO cells using recombinant DNA technology. After recovery from the cell culture media, the protein is subjected to a number of high-resolution chromatography and filtration steps. The purified protein is lyophilized in the presence of sucrose, glycine, L-glutamic acid, sodium chloride, and polysorbate 80 (as excipients).

The absorbable collagen sponge is the matrix for the rhBMP-2. It consists of bovine type I collagen from the deep flexor tendon. The resuspended rhBMP-2 applied on the absorbable collagen sponge constitutes the InFUSE Bone Graft, which is then introduced into the LT-CAGE Lumbar Tapered Fusion Device. This fusion divice consists of a metallic cylindric tapered cage made of implant-grade titanium alloy, which is filled with the bone graft compound and implanted via an open anterior surgical approach or a laparoscopic anterior surgical approach. The LT-CAGE Lumbar Tapered Fusion Device (marketed outside the U.S. since 1995 and distributed in the U.S. since 1999) replaces the previously approved spinal fusion cage INTEX FIX Threaded Fusion Device.

Overview of Therapeutic Properties

The InFUSE Bone Graft/LT-CAGE Lumbar Tapered Fusion Device is used in spinal fusion procedures in patients suffering degenerative disc disease at the lower region of the spine from L4-S1. It is implanted via an anterior open or an anterior laparoscopic approach in skeletally mature patients who have had at least 6 months of nonoperative treatment before the implantation. Patients might also have up to grade I spondylolisthesis. The bone graft component of the product, which stimulates new bone formation at the site of implantation, is held in place within the metallic cage, which stabilizes and maintains the spacing in the spinal region.

The use of the InFUSE Bone Graft/LT-CAGE Device was found to be as effective as the use of the cage filled with autograft bone, with the advantage of reduced morbidity and pain for patients. It exhibited a 68% chance of overall success 24 months from the surgery in the case of the anterior laparoscopic surgical approach and 57% in the case of the open surgical approach, with the latter comparable to success of the autograft procedure.

Most of the observed side effects after the use of the InFUSE Bone Graft/LT-CAGE Device were similar in frequency and severity to those observed with the use of the autograft bone procedure. Urogenital events and retrograde ejaculation rates were more frequent in the case of the InFUSE Bone Graft/LT-CAGE Device than in the autograft procedure.

The development of anti-rhBMP-2 antibodies has been observed only rarely in patients, while 18% of patients developed antibovine type I collagen antibodies. No studies have been carried out on patients with hepatic or renal impairment, metabolic bone diseases, autoimmune or immunosuppressive diseases, or on patients 65 or older. The use of the InFUSE Bone Graft/LT-CAGE Device is contraindicated in patients with an infection, a tumor, a removed tumor at the site of implantation, and during pregnancy and lactation.

Further Reading

http://www.fda.gov

http://www.medtronic.com

Boden, S.D. et al., The use of rhBMP-2 in interbody fusion cages. Definitive evidence of osteoinduction in humans: a preliminary report, *Spine*, 25, 376–81, 2000.

Burkus, J.K. et al., Clinical and radiographic outcomes of anterior lumbar interbody fusion using recombinant human bone morphogenetic protein-2, *Spine*, 27, 396–408, 2002.

Burkus, J.K. et al., Radiographic assessment of interbody fusion using recombinant human bone morphogenetic protein type 2, *Spine*, 28, 372–377, 2003.

Insuman

Product Name: Insuman (trade name)

Insulin human (international nonproprietary name)

Description: Insuman is a recombinant insulin, produced in *Escherichia coli,* which is identical to native human insulin. Insuman is provided, mainly for subcutaneous administration, in the following forms (generally at strengths of either 40 or 100 IU/ml): Insuman Rapid, a solution of unmodified insulin; Insuman Basal, a suspension of insulin in crystalline form; Insuman Comb 15, a combined formulation of 15% soluble insulin and 85% insulin crystals; Insuman Comb 25, containing 25% soluble insulin and 75% insulin crystals; Insuman Comb 50, containing 50% soluble insulin and 50% insulin crystals; and Insuman Infusat, a solution of insulin to be used for pump infusion. Insuman is also provided in a prefilled multidose device, the OptiSet.

Approval Date: 1998 (E.U.). This product replaces a previous identical formulation (but different manufacturing process) approved in 1997, which is no longer being marketed.

Therapeutic Indication: Insuman is indicated for the treatment of diabetes mellitus in order to provide blood glucose homeostasis.

Manufacturer: Aventis Pharma Deutchland GmbH, D-65926, Frankfurt, Main, Germany

Marketing: Aventis Pharma Deutchland GmbH, D-65926, Frankfurt, Main, Germany

Manufacturing

Insuman is a recombinant human insulin identical to the natural molecule. It is produced by recombinant DNA technology in a modified *E. coli* strain as a fusion protein with a bacterial peptide attached. The protein is processed after production to remove the bacterial component, rendering it identical to the human molecule. Extensive chromatographic purification of the product is carried out. Insuman is provided in a number of forms. Insuman Rapid is a soluble formulation containing unmodified human insulin, m-cresol (as a preservative), sodium dihydrogen phosphate dihydrate (as a buffering agent), glycerol (as an isotonic agent), hydrochloric acid (to dissolve the insulin), and sodium hydroxide (for pH adjustment). Insuman Basal is a suspension of insulin in a crystallized form, which has a slower onset of action and contains human insulin, protamine sulphate (as a crystallizing agent), m-cresol and phenol (as preservatives), zinc chloride (a promoter of complex formation), sodium dihydrogen phosphate dihydrate (as a buffering agent), glycerol, hydrochloric acid, and sodium hydroxide. Insuman Comb is a combination of the soluble and the crystalline forms of insulin and is provided in three different formulations: Insuman Comb 15, Insuman Comb 25, and Insuman Comb 50, containing 15, 25, and 50%, respectively, of soluble insulin with the remainder in crystallized form. The three preparations contain human insulin, protamine sulphate, m-cresol, phenol, zinc chloride, sodium dihydrogen phosphate dihydrate, glycerol, hydrochloric acid, and sodium hydroxide. Insuman Infusat is a solution for use with insulin pumps and contains dissolved human insulin, phenol, zinc chloride (as a stabilizer), trometamol (as a buffering agent), poloxamer 171 (as stabilizing agent), glycerol, and hydrochloric acid. Insuman Rapid, Insuman Basal, and Insuman Comb are each provided in a prefilled multidose device, OptiSet.

The shelf life of the products is 24 months when stored at 2 to 8°C, protected from direct light. Quality tests and extensive characterization of the molecule is undertaken to ensure quality and safety of the final product. Tests include mass spectroscopy, protein sequencing, H-NMR and assays for the presence of bacterial contaminants, and endotoxins.

Overview of Therapeutic Properties

Insuman supercedes the semi-synthetic insulin of porcine origin and the previously marketed human insulin HGT, which was identical to Insuman in its end product but was produced using a different manufacturing process. Insuman is used to maintain a constant level of glucose in the blood of diabetes mellitus patients. The different formulations have different onsets of action and duration. Insuman Rapid is an unmodified insulin and should be administered 15 to 20 minutes before meals, while Insuman Basal has a slower absorption and should be administered 45 to 60 minutes before meals. Insuman Comb 50 should be administered 20 to 30 minutes before meals, and Insuman Comb 15 and Insuman Comb 25 should be administered roughly 30 to 45 minutes before meals. Administration should be subcutaneous, except in rare occasions where intramuscular administration is used, with rotation of the site of injections within the same area. Dosages are adjusted to meet individual needs of patients. Care should be taken with coadministration of drugs that affect glucose homeostasis in the blood due to possible interference with efficacy of Insuman.

Insuman Rapid is also used as an intravenous injection in the treatment of hyperglycemic coma and ketoacidosis, and for stabilization of diabetes mellitus patients undergoing surgery.

Studies proved that Insuman is as effective as other human insulin preparations, and the observed side effects, such as hypoglycemia, infections, temporary visual disorders, and lipohypertrophy at the site of injection, were comparable to other insulin products. Care should be taken in the use of Insuman for patients with diabetic retinopathy. Insuman is contraindicated in the case of hypoglycemia.

Further Reading

http://www.aventis.com

http://www.eudra.org

Johansson, U.B. et al., Improved glucose variability HBAlc insuman infusat and less insulin in IDDM patients using insulin lispro in CS II. The Swedish multicenter lispro insulin study, *Diabetes Metab.*, 26(5), 423–423, 2000.

Selam, J.H., External and implantable insulin pumps: current practice in the treatment of diabetes, *Exp. Clin. Endocr. Diab.*, 109, S333–S340, 2001.

Intron A

Product Name: Intron A (trade name)

Interferon alfa-2b (international nonproprietary name)

Description: Intron A is a recombinant human interferon alfa-2b (see also the monograph for the product Viraferon). The 19.3-kDa, single-chain nonglycosylated polypeptide is produced by recombinant DNA technology in *Escherichia coli*. It is presented in a lyophilized form (at strengths up to 30 million IU activity) containing human serum albumin and as an albumin-free solution in vials and cartridges to be used with multidose devices for subcutaneous or intravenous administration.

Approval Date: 1986 (U.S.); 2000 (E.U.)

Therapeutic Indications: Intron A is indicated for the treatment of chronic hepatitis B and C, hairy cell leukemia, chronic myelogenous leukemia, multiple myeloma follicular non-Hodgkin's lymphoma, carcinoid tumor, malignant melanoma, condylomata acuminata, and Kaposi's sarcoma.

Manufacturer: SP (Brinny) Company, Innishannon, County Cork, Ireland, http://www.schering-plough.com (manufacturer is responsible for batch release in the European Economic Area)
Schering Corporation, Galloping Hill Road, Kenilworth, NJ 07033, http://www.schering.com (U.S.)

Marketing: SP Europe, 73 rue de Stalle, 1180 Bruxelles, Belgium, http://www.schering-plough.com (E.U.)
Schering Corporation, Galloping Hill Road, Kenilworth, NJ 07033, http://www.schering.com (U.S.)

Manufacturing

Intron A is a recombinant human interferon alfa-2b. The gene-encoding human interferon alfa-2b was derived from human leukocytes and cloned into a plasmid, from which the protein is expressed in a modified *E. coli* strain. Purification, subsequent to extraction, involves various chromatographic steps, procedures to remove viral and microbial contaminants, crystallization, and resuspension. Intron A is provided either in a lyophilized form with a solvent for reconstitution or as a solution in vials and cartridges to be used with multidose devices. The lyophilized formulation contains interferon alfa-2b (active), as well as glycine, dibasic sodium phosphate, monobasic sodium phosphate, and human serum albumin as excipients. When provided as a solution, Intron A contains interferon alfa-2b (active), and dibasic sodium phosphate, monobasic sodium phosphate, disodium edetate, sodium chloride, m-cresol, and polysorbate 80 as excipients.

The shelf life of the products differs for the various formulations but ranges from 15 months to 3 years when stored at 2 to 8°C. The quality and the safety of the product are ensured by validation tests that include SDS-PAGE, IEF, HPLC, MTT-CPE, and LAL assay.

Overview of Therapeutic Properties

Interferons are molecules involved in the defense against viral infections. Interferon alfa-2b promotes inhibition of viral replication inside infected cells, and it has an immunomodulatory activity that leads to induction of phagocytosis in macrophages and cytotoxicity in lymphocytes. It also suppresses cell proliferation.

Interferon alfa-2b is indicated for the treatment of numerous clinical conditions. It is used in the treatment of chronic hepatitis B when there is evidence of viral replication, elevated ALT, and active liver inflammation and fibrosis. Administration is subcutaneous, three times a week, for up to 6 months, at a dosage matched with individual needs. Intron A is also indicated for the treatment of chronic hepatitis C when there is evidence of viral replication. In this case it is administered three times a week for up to 18 months. Intron A proved to be more efficient in the treatment of hepatitis C when administered in combination with ribavirin.

Intron A is used for the treatment of hairy cell leukemia, which is characterized by the presence of circulating B-lymphocytes with cytoplasmatic projections, which infiltrate bone marrow and spleen. The treatment involves subcutaneous administration of Intron A three times a week, leading to normalization of some hematological values within 1 or 2 months, and normalization of all three values within 6 months in 80% of patients. A partial response is achieved in 60% of patients and a complete response in 11%.

Daily subcutaneous administration of IntronA is used for the treatment of chronic myelogenous leukemia. A combination therapy with cytarabine proved more successful than monotherapy with Intron A, with greater reduction of tumor burden and an increased survival after 3 years. Intron A is used as a maintenance therapeutic after chemotherapy in patients with multiple myeloma. Subcutaneous administration of IntronA three times a week proved to decrease the progression of the disease, but data did not indicate an increased survival. Intron A is given subcutaneously three times a week, for up to 18 months, to patients with follicular non-Hodgkin's lymphoma as an adjuvant to chemotherapy and has been shown to extend survival in these patients. Subcutaneous administration of Intron A three times a week in patients with carcinoid tumors, which are found mainly in the lung or in the gastrointestinal tract, resulted in a response (remission of diarrhea and flushing) in 70% of patients, while regression of the tumor was observed in a small number of patients.

Administration of Intron A in postoperative patients with a high risk of recurrence of malignant melanoma resulted in an increased rate of survival. Intron A is given intravenously daily 5 days a week for 4 weeks during the induction therapy, and subcutaneously three times a week for 48 weeks afterward as a maintenance treatment. Intron A is used to treat condylomata acuminata, a form of genital warts caused by a human papillomavirus, in combination with podophyllum treatment. Administration of Intron A intralesionally three times a week for 3 weeks should lead to a response after 4 to 6 weeks; a second treatment might be required. Intron A is also used in the treatment of AIDS-related Kaposi's sarcoma. In this case, optimal responses were achieved in patients with no systemic symptoms.

The most common side effects observed upon administration of Intron A were flu-like symptoms, loss of appetite, nausea, and variation in hematic and enzymatic values. Very rarely, depressive and suicidal behavior was reported. Intron A is contraindicated during pregnancy and lactation.

Further Reading

http://www.eudra.org
http://www.fda.gov
http://www.introna.com
http://www.schering-plough.com

Basser, R.L. et al., Recombinant alpha-2b interferon in patients with malignant carcinoid tumour, *Aust. N. Z. J. Med.*, 21, 875–878, 1991.

Blade, J. et al., Maintenance treatment with interferon alpha-2b in multiple myeloma: a prospective randomized study from PETHEMA (Program for the Study and Treatment of Hematological Malignancies, Spanish Society of Hematology), *Leukemia*, 12, 1144–1148, 1998.

Bourantas, K.L. et al., Prolonged interferon-alpha-2b treatment of hairy cell leukemia patients, *Eur. J. Haematol.*, 64, 350–351, 2000.

Gisslinger, H., Interferon alpha in the therapy of multiple myeloma, *Leukemia,* 11(Suppl. 5), S52–S56, 1997.

Janssen, H.L. et al., Interferon alfa for chronic hepatitis B infection: increased efficacy of prolonged treatment. The European Concerted Action on Viral Hepatitis (EUROHEP), *Hepatology,* 30, 238–243, 1999.

Kirkwood, J.M. et al., Immunomodulatory effects of high-dose and low-dose interferon alpha2b in patients with high-risk resected melanoma: the E2690 laboratory corollary of intergroup adjuvant trial E1690, *Cancer,* 95, 1101–1112, 2002.

Maloisel, F. et al., Results of a phase II trial of a combination of oral cytarabine ocfosfate (YNK01) and interferon alpha-2b for the treatment of chronic myelogenous leukemia patients in chronic phase, *Leukemia,* 16, 573–580, 2002.

Neri, N. et al., Chemotherapy plus interferon-alpha2b versus chemotherapy in the treatment of follicular lymphoma, *J. Hematother. Stem. Cell. Res.*, 10, 669–674, 2001.

Rybojad, M. et al., Non-AIDS-associated Kaposi's sarcoma (classical and endemic African types): treatment with low doses of recombinant interferon-alpha, *J. Invest. Dermatol.*, 95(Suppl. 6), S176–S179, 1990.

Saracco, G. et al., A randomized 4-arm multicenter study of interferon alfa-2b plus ribavirin in the treatment of patients with chronic hepatitis C relapsing after interferon monotherapy, *Hepatology,* 36, 959–966, 2002.

Schober, C. et al., Antitumour effect and symptomatic control with interferon alpha 2b in patients with endocrine active tumours, *Eur. J. Cancer,* 28A, 1664–1666, 1992.

Solal-Celigny, P. et al., Recombinant interferon alfa-2b combined with a regimen containing doxorubicin in patients with advanced follicular lymphoma. Groupe d'Etude des Lymphomes de l'Adulte . *N. Engl. J. Med.*, 329, 1608–1614, 1993.

Verbaan, H.P. et al., High sustained response rate in patients with histologically mild (low grade and stage) chronic hepatitis C infection. A randomized, double blind, placebo controlled trial of interferon alpha-2b with and without ribavirin, *Eur. J. Gastroenterol. Hepatol.*, 14, 627–633, 2002.

Zee, B. et al., Quality-adjusted time without symptoms or toxicity analysis of interferon maintenance in multiple myeloma, *J. Clin. Oncol.*, 16, 2834-2839, 1998.

Kineret

Product Name:	Kineret (trade name) Anakinra (international nonproprietary name)
Description:	Anakinra is a recombinant form of the human interleukin-1 receptor antagonist. The 17.3-kDa, 153-amino acid single chain polypeptide differs from the native molecule only in containing an additional N-terminal methionine residue. It is produced by recombinant DNA technology in *Escherichia coli* cells and is provided as a solution (containing 100 mg product in a volume of 0.67 ml) for subcutaneous injection.
Approval Date:	2001 (U.S.); 2002 (E.U.)
Therapeutic Indications:	Kineret is indicated for the treatment of rheumatoid arthritis, in combination with methotrexate, in adults who do not respond to methotrexate alone.
Manufacturer:	Amgen Europe B.V., Minervum 7061, 4817 ZK Breda, the Netherlands, http://www.amgen.com (manufacturer is responsible for batch release in the European Economic Area) Amgen, Inc., One Amgen Center Drive, Thousand Oaks, CA 91329-1789, http://www.amgen.com (U.S.)
Marketing:	Amgen Europe B.V., Minervum 7061, 4817 ZK Breda, the Netherlands (E.U.) Amgen, Inc., One Amgen Center Drive, Thousand Oaks, CA (U.S.)

Manufacturing

Anakinra is a recombinant human interleukin-1 receptor antagonist. It is produced by recombinant DNA technology in *E. coli*. The protein is purified from *E. coli* extract using several chromatographic steps and filtration procedures. The final product consists of the active substance anakinra, as well as sodium citrate, sodium chloride, disodium edetate, polysorbate 80, and sodium hydroxide as excipients.

The shelf life of the product is 18 months when stored at 2 to 8°C, protected from direct light. Extensive tests are carried out to ensure the quality and safety of the product.

Overview of Therapeutic Properties

Kineret is an interleukin-1 receptor antagonist. It binds interleukin-1α and interleukin-1β cell surface receptors but without inducing a biological response. The product therefore effectively blocks interleukin-1 activity, which is considered a critical mediator of the inflammation and joint damage characteristic of rheumatoid arthritis. Kineret is indicated in the treatment of rheumatoid arthritis in adult patients who have not responded to treatment with methotrexate alone. Kineret should be administered subcutaneously daily in combination with methotrexate.

Kineret proved to be effective in the treatment of rheumatoid arthritis leading to a reduction of signs and symptoms of the disease a few weeks after initiating the treatment. The most common side effects observed with Kineret were mild reactions at the injection site and headaches. Neutropenia was reported in some cases, as well as allergic reactions and severe infections. Neutralizing antibodies have been rarely observed but with no effect on the efficacy of the treatment. Kineret is contraindicated during pregnancy and lactation.

Further Reading

http://www.amgen.com
http://www.eudra.org
http://www.fda.gov
http://www.kineretrx.com
Cvetkovic, R.S. and Keating, G., Anakinra, *BioDrugs*, 16, 303–311, 2002.
Nuki, G. et al., Long-term safety and maintenance of clinical improvement following treatment with anakinra (recombinant human interleukin-1 receptor antagonist) in patients with rheumatoid arthritis: extension phase of a randomized, double-blind, placebo-controlled trial, *Arthritis Rheum.*, 46, 2838–2846, 2002.

Lantus

Product Name:	Lantus (trade name) Insulin glargine (international nonproprietary name)
Description:	Lantus is a long-acting human insulin analogue (and appears to be identical to the product Optisulin). Lantus differs from natural human insulin at the C-terminal end of the B chain, which has two additional arginine residues, and at the C-terminus of the A chain, where asparagine residue number 21 is replaced by a glycine. Its overall molecular mass is 6,063 Da. Lantus is produced in modified *Escherichia coli* by recombinant DNA technology and is presented as a solution (100 IU/ml) for subcutaneous injection.
Approval Date:	2000 (E.U. and U.S.)
Therapeutic Indications:	Lantus is indicated for patients with diabetes mellitus where insulin treatment is required.
Manufacturer:	Aventis Pharma Deutschland GmbH, Brünistrasse 50, 65926 Frankfurt am Main, Germany, http://www.aventis.com (E.U. and U.S.)
Marketing:	Aventis Pharma Deutschland GmbH, Brünistrasse 50, 65926 Frankfurt am Main, Germany (E.U.) Aventis Pharmaceuticals, Inc., Kansas City, MO 64137 (U.S.)

Manufacturing

Insulin glargine is a recombinant human insulin analogue. It is produced by recombinant DNA technology in *E. coli*, following cloning of the modified human insulin chains. The purification process involves a number of chromatographic steps followed by precipitation. The final product contains the active ingredient insulin glargine and the following excipients: zinc-chloride (as a stabilizer), m-cresol (as a preservative), glycerol (as a tonicity agent), hydrochloric acid (to dissolve the active ingredient), and sodium hydroxide (to adjust the pH). It is presented as a solution with a pH of 4 in vials or cartridges to be used with multidose devices for subcutaneous injection.

The shelf life of the product is 24 months when stored at 2 to 8°C. Extensive tests have been carried out to ensure the quality and the safety of the product.

Overview of Therapeutic Properties

Lantus is a long-acting human insulin analogue. The genetically modified molecule has an isoelectric point that differs from that of natural human insulin, resulting in a molecule that is more soluble in an acid environment and less stable at physiological pH. When the product is injected into a pH-neutral environment, it forms microprecipitates from which insulin is released slowly over a long period of time. Administration of Lantus is not followed by the usual activity peak that is observed with native human insulin, and its action continues at a constant rate for up to 24 hours. Lantus is indicated for patients with diabetes mellitus, where insulin is required. The dosage should be adjusted for the individual needs of the patient, and it should be administered once daily, in the evening, with rotation of the injection site.

Lantus proved its efficacy in controlling blood glucose levels in patients with type 1 and type 2 diabetes mellitus. The most common observed side effects are those commonly reported for insulin and other insulin analogues, such as hypoglycemia, immunoreactions, development of antibodies, and visual disorders. Pain and reactions at the site of injections are observed more pronouncedly when Lantus is used compared to other insulin products. No studies have been carried out on children. Lantus is contraindicated in the case of hypoglycemia. It should never be mixed with other products.

Further Reading

http://www.aventis.com
http://www.eudra.org
http://www.fda.gov
http://www.lantus.com

McKeage, K. and Goa, K.L., Insulin glargine: a review of its therapeutic use as a long-acting agent for the management of type 1 and 2 diabetes mellitus, *Drugs*, 61, 1599–1624, 2001.

Owens, D.R. and Griffiths, S., Insulin glargine (Lantus), *Int. J. Clin. Pract.*, 56, 460–466, 2002.

Leukine

Product Name: Leukine (trade name)

Sargramostim (common name)

Description: Sargramostim is a recombinant human granulocyte-macrophage colony stimulating factor (rhuGM-CSF), a 127-amino acid glycosylated hematopoietic growth factor. It differs from the native human molecule by a substitution of leucine at position 23. It is presented in both liquid (500 μg/ml) and lyophilized (250 μg) formats.

Approval Date: 1991 (U.S.)

Therapeutic Indications: Leukine was originally indicated for use following induction of chemotherapy in adult patients with acute myelogenous leukemia (AML), also known as acute nonlymphocytic leukemia (ALL), in order to shorten time to neutrophil recovery and reduce the incidence of severe infection. It is also indicated for use in mobilization, following transplantation of autologous peripheral blood progenitor cells, for use in myeloid reconstitution following autologous and allogenic bone transplantation, and for use in bone marrow transplantation failure or engraftment delay.

Manufacturer: Berlex Laboratories Inc., Richmond, CA 94804, http://www.berlex.com

Marketing: Berlex Laboratories Inc., Richmond, CA 94804

Manufacturing

Sargramostim is produced in a recombinant yeast (*Saccharomyces Cerevisiae*) expression system. Subsequent to fermentation, the yeast cell mass is removed from the product-containing extracellular liquid by filtration. After a concentration step (ultrafiltration), the product is purified using a series of chromatographic steps, including ion exchange and preparative HPLC. In addition to the active substance, Leukine also contains mannitol, sucrose, and tromethamine as excipients. The final product undergoes extensive testing, including electrophoretic and HPLC analysis.

Overview of Therapeutic Properties

GM-CSF, also known as CSF-α or pluripoietin-α, is a colony-stimulating factor that promotes the proliferation and differentiation of hematopoietic progenitor cells, particularly those yielding neutrophils and macrophages, but also eosinophils, erythrocytes, and megakaryocytes. It also promotes the activation of mature hematopoietic cells, resulting in enhanced phagocytic activity, enhanced microbiocidal activity, and enhanced leukocyte chemotaxis. It brings about its effect by binding a heterodimeric GM-CSF receptor present on the surface of susceptible cells.

In older patients suffering from ALL, administration of Leukine significantly accelerated neutrophil recovery and reduced the incidence of severe and life-threatening infections. The product is generally administered daily, intravenously, over a 4-hour period, usually starting approximately on day 11 of the chemotherapy regime. The safety and efficacy of Leukine has not been assessed in AML patients under 55.

In transplantation patients, Leukine increases white blood cells after autologous and allogeneic bone marrow transplantation, as well as after peripheral blood progenitor cell (PBPC) transplantation. When Leukine was administered after transplantation with mobilized PBPCs, there was a further acceleration in white blood cell recovery. Additionally, the number of platelet transfusions and red blood cell transfusions necessary decreased, and subjects were discharged from the hospital earlier when compared to those not receiving Leukine.

Although generally well tolerated, edema, capillary leak syndrome, and pleural and pericardial effusion have been reported in patients following Leukine administration. Respiratory and cardiovascular symptoms have also been noted. Leukine is contraindicated in patients with excessive leukemic myeloid blasts in the bone marrow or peripheral blood, in patients with known hypersensitivity to GM-CSF, and for concomitant use with chemotherapy or radiotherapy.

Further Reading

Anon., Leukine maintains viral suppression and extends duration of anti- retroviral therapy, *Aids Patient Care St.*, 13(9), 568–569, 1999.

McLaughlin, P. et al., Rituximab in combination with GM-CSF (leukine) for patients with recurrant indolent lymphoma, *Blood*, 98(11), 2536, 2001.

Nemunaitis, J. et al., Retrospective analysis of infectious disease in patients who received leukine versus patients not receiving a cytokine who underwent autologous bone marrow transplant (ABMT) for treatment of lymphoid cancer, *Blood*, 86 (10), 871–871, 1995.

Spitler, L.E., Adjuvant therapy of melanoma, *Oncology-NY*, 16(1), 40–48, 2002.

LeukoScan

Product Name: LeukoScan (trade name)

Sulesomab (international nonproprietary name)

Description: LeukoScan consists of the antigen-binding Fab' fragments of the monoclonal antibody sulesomab. Sulesomab displays binding affinity for the surface granulocyte nonspecific cross-reacting antigen (NCA90), present on the surface of granulocytes, and also for the carcinoembryonic antigen (CEA). It is produced in mice by ascites technology and, after processing, is provided in a lyophilized form (0.31 mg/vial). Before administration as an intravenous injection, LeukoScan is conjugated with the radioactive technetium-99m isotope (^{99m}Tc, obtained separately) for immunoscintigraphy.

Approval Date: 1997 (E.U.)

Therapeutic Indications: LeukoScan is used in diagnostic imaging to target infection and inflammation loci in long bones in patients with suspected osteomyelitis and with diabetic foot ulcers.

Manufacturer: Eli Lilly Pharma Fertigung und Distribution GmbH & Co. KG., Teichweg 3, 35396 Giessen, Germany, http://www.lilly.com (manufacturer is responsible for import and batch release in European Economic Area)

Marketing: Immunomedics Europe, Haarlemmerstraat 30, 2181 HC Hillegom, the Netherlands, http://www.immunomedics.com

Manufacturing

The hybridoma cell line producing the sulesomab monoclonal antibody was obtained by fusion of lymphocytes from CEA immunized mice and the SP2/0 mouse myeloma cell line. The hybridoma cells are injected into pristane-primed mice, and 14 to 28 days after inoculum ascites are removed from the mice by tapping. The sulesomab IgG is then purified from the ascites fluid by combined chromatographic steps and processed by digestion with pepsin to produce $F(ab')_2$ fragments. This is followed by their chemical reduction to individual Fab'-SH fragments using cysteine. The final product is presented in a lyophilized form, composed mainly of the sulesomab Fab'-SH but also containing $F(ab')_2$ and H and L chain fragments and the following excipients: stannous chloride (as a reducing agent), sodium potassium tartrate and sodium acetate (as weak chelating agents preventing the precipitation of stannous chloride at physiological pH), sodium chloride and sucrose (as stabilizers), argon and traces of acetic acid, and hydrochloric acid for a final buffered solution at pH 5 to 7.

The lyophilized form of the product has a shelf life of 48 months when stored at 2 to 8°C. Routine evaluative tests were carried out to verify the purity of the final product and the quality of the intermediates (HPLC, IEF, GC, SDS-PAGE, and protein A assay), the potency of the product (immunoreactivity and flow cytometric assays), and microbiological and viral safety (mycoplasma assay, amplified XC plaque, MAP, amplified S^+L^-focus, and reverse transcriptase). Routine tests were carried out to verify endotoxin and pyrogen tests.

Overview of Therapeutic Properties

The difficulty in detecting infection and inflammation at early stages in bones, or after surgery, necessitates the use of radionucleotides. LeukoScan, which binds to granulocytes at the site of infection, offers advantages of rapid localization, simple use, and no handling of patient blood. In clinical trials of patients with long bone osteomyelitis and diabetic foot ulcers, imaging with LeukoScan yielded a statistically significant increase in sensitivity (95 vs. 84%) and accuracy (76.6 vs. 70.9%) compared to white blood cell imaging with *in vitro* labeling of autologous leukocytes from patients. The most common side effect, observed 24 hours after administration of LeukoScan, was a transient reduction in leukocyte and neutrophil counts (within the normal range) that was found to be of no clinical significance.

The use of the Fab' fragment instead of the monoclonal antibody reduces immunotoxicity (no activation of complement, no crosslinking of antigen on cell surface) and immunogenicity (negligible human antimurine antibody [HAMA] response), while there is no detectable pharmacological effect on granulocyte function. The monovalent Fab' fragment is rapidly cleared from the blood, allowing imaging to be carried out shortly after injection. Finally, the reduced form of the Fab' exposes free thiol groups, which are used for direct radiolabeling with the ^{99m}Tc isotope.

^{99m}Tc emits radiation of a relatively low energy that is detectable by a standard gamma camera. It has advantages over other radioisotopes such as ^{67}Ga, which requires a high radiation dose and a long interval between administration and imaging, and ^{123}I and ^{111}In, which have longer half-lives, leading to increased exposure to the patient. Radiation doses to organs are within the accepted range for radiopharmaceuticals.

LeukoScan has recently been shown to permit rapid and accurate diagnosis of acute, nonclassic appendicitis and subacute endocarditis, indicating its potential use in detection of inflammatory bowel disease, pelvic inflammatory disease, and fever of unknown origin.

LeukoScan is contraindicated during pregnancy because of the risks of radiation exposure to the fetus. Cross-reaction with the CEA antigen could result in interaction with CEA-expressing tumors.

Further Reading

http://www.eudra.org

http://www.immunomedics.com

http://www.leukoscan.com

Becker, W. et al., Detection of soft-tissue infections and osteomyelitis using a technetium-99m-labeled anti-granulocyte monoclonal antibody fragment, *J. Nucl. Med.*, 35, 1436–1443, 1994.

Becker, W. et al., Rapid imaging of infections with a monoclonal antibody fragment (LeukoScan), *Clin. Orthop.*, 329, 263–272, 1996.

Schroeter, S. and Greiner-Bechert, L., LeukoScan protocol, *Nucl. Med. Commun.*, 22, 841, 2001.

Liprolog (withdrawn from market)

Product Name: Liprolog (trade name)

Insulin lispro (international nonproprietary name)

Description: Liprolog is a fast-acting insulin analogue (see also the monograph for the product Humalog). It differs from the human insulin in an inversion of the natural proline-lysine sequence at positions 28 and 29 of the insulin B chain. It is produced by recombinant DNA technology in a modified *Escherichia coli* strain. The final product is presented as a solution for subcutaneous injection in vials and cartridges for multidose devices (Liprolog pen). Formulations containing 25% (Liprolog Mix 25) or 50% (Liprolog Mix 50) of insulin lispro in soluble form and 75 or 50%, respectively, of insulin lispro crystallized with protamine sulphate were also available.

Approval Date: 1997 (E.U.)

Withdrawal Date: 2001

Therapeutic Indications: Liprolog was used to promote normal glucose homeostasis in patients suffering from diabetes mellitus, as well as for the initial stabilization of the disease.

Manufacturer: Lilly France S.A., Rue du Colonel Lilly, 67640 Fegersheim, France; Lilly Pharma Fertigung und Distribution GmbH and Co. KG, Teichweg 3, 35396 Giessen, Germany; and Lilly S.A., Avda. de la Industria 30, 28108 Alcobendas (Madrid), Spain, http://www.lilly.com (manufacturers are responsible for import and batch release in the European Economic Area)

Marketing: Eli Lilly Nederland B.V., Grootslag 1-5, 3991 RA Houten, the Netherlands http://www.lilly.com

Luveris

Product Name: Luveris (trade name)

Lutropin alfa (international nonproprietary name)

Description: Lutropin alfa is a recombinant human luteinizing hormone (r-hLH) produced in Chinese hamster ovary (CHO) cells using recombinant DNA technology. The product is a hereterodimeric glycoprotein, consisting of a 14-kDa, 92-amino acid α-subunit and a 15-kDa, 121-amino acid β-subunit. Luveris is presented as a lyophilized powder (3.4 μg, equivalent to 75 IU activity/vial), to be reconstituted using the provided solvent prior to subcutaneous administration.

Approval Date: 2000 (E.U.)

Therapeutic Indications: Luveris is indicated for the stimulation of follicular development in women suffering from severe luteinizing hormone (LH) and follicular stimulating hormone (FSH) deficiency. Luveris is used in combination with FSH administration.

Manufacturer: Industria Farmaceutica Serono S.p.A., Bari, Italy, http://www.serono.com (manufacturer is responsible for batch release in the European Economic Area)

Marketing: Serono Europe Limited, 56 Marsh Wall, London E14 9TP, U.K., http://www.serono.com (E.U.)

Manufacturing

The genes encoding the α- and β-chains of the human luteinizing hormone were cloned into a modified CHO cell line. After recovery from the cell culture media, the product is concentrated by ultrafiltration and purified using a number of high-resolution chromatographic steps. The final product consists of a reproducible mixture of isoforms, the active substance lutropin alfa, and the following excipients: polysorbate 20, monohydrate sodium dehydrogen phosphate, dehydrate disodium phosphate, phosphoric acid, sodium hydroxide, and nitrogen.

The shelf life of the product is 24 months when stored at a temperature below 25°C. Extensive tests carried out to ensure quality and safety during production and on the final product include amino acid sequencing, peptide mapping, chromatography, mass spectrometry, ELISA, and *in vivo* bioassays.

Overview of Therapeutic Properties

Hormone replacement therapy is considered for women suffering from hypogonadotrophic hypogonadism, where anovulation is due to LH and FSH deficiency. Administration of human menopausal gonadotrophins isolated from urine faces problems with purification and limited supply, leading to its replacement by r-hLH and recombinant human FSH. The r-hLH exhibits an activity comparable to the natural hormone and stimulates the production of oestradiol, which supports the development of the follicle stimulated by FSH. Luveris should be administered daily as a subcutaneous injection for up to 3 weeks, or up to 5 weeks if required, at a dosage adjusted to individual needs, and in combination with FSH. When the required response has been achieved, an injection of human chorionic gonadotrophin (hCG) should be administered 24 to 48 hours after the last dose of Luveris and FSH to stimulate the release of the mature egg from the follicle.

Clinical studies were performed on a small number of women, due to the rarity of the condition, and showed the efficacy of Luveris in inducing ovulation when administered in combination with recombinant human FSH and hCG.

The most commonly reported side effects were headaches, abdominal and breast pain, nausea, ovarian cysts, and reactions at the site of injection. Monitoring of the ovary should take place during the treatment in order to prevent excessive ovarian response, which may lead to ovarian hyperstimulation syndrome (OHSS). Ectopic pregnancies and a high incidence of multiple pregnancies have been reported. Luveris should not be administered during pregnancy and lactation.

Further Reading

http://www.eudra.org
http://www.serono.com

Burgues, S., The effectiveness and safety of recombinant human LH to support follicular development induced by recombinant human FSH in WHO group I anovulation: evidence from a multicentre study in Spain, *Hum. Reprod.*, 16, 2525–2532, 2001.

Lisi, F. et al., Use of recombinant LH in a group of unselected IVF patients, *Reprod. Biomed. Online*, 5, 104–108, 2002.

LYMErix (withdrawn from market)

Product Name:	LYMErix (trade name) Lyme disease vaccine (recombinant OspA) (international nonproprietary name)
Description:	LYMErix is a vaccine consisting of a recombinant outer surface lipoprotein A (L-OspA) from *Borrelia burgdorferi sensu stricto*. The 257-amino acid lipoprotein is produced in *Escherichia coli* and is provided as a suspension for intramuscular injection. Each single dose contains 30 μg active substance in 0.5 ml.
Approval Date:	1998 (U.S.)
Withdrawal Date:	2002 (due to commercial reasons)
Therapeutic Indications:	LYMErix was indicated for immunization against Lyme disease in individuals 15 to 70 years old.
Manufacturer:	GlaxoSmithKline Biologicals S.A., Rue de l'institut 89, 1330 Rixensart, Belgium, http://www.gsk.com
Marketing:	GlaxoSmithKline Biologicals S.A., Rue de l'institut 89, 1330 Rixensart, Belgium, http://www.gsk.com

Manufacturing

The gene encoding the outer surface lipoprotein A from the bacterium *B. burgdorferi sensu stricto* has been cloned and transformed into a modified *E. coli* strain. The full-length protein is produced in the bacterial culture, and its associated lipid component is attached within the cell after translation. The lipoprotein is purified from the culture and adsorbed onto aluminium hydroxide. The final product consists of a liquid formulation for intramuscular administration and contains the lipoprotein OspA, as well as aluminium hydroxide, phosphate buffered saline, and 2-phenoxyethanol as excipients.

Overview of Therapeutic Properties

Lyme disease is the most common vector-borne disease in North America, with an increasing incidence rate over the past several decades. *B. burgdorferi* is the bacterial causative agent and is transmitted to humans through the bite of an infected tick. The disease has a bimodal age distribution, with a high incidence in individuals 2 to 15 and 30 to 55 years old. The disease develops in three stages: the first stage consists of a rash, with flu-like or meningitis symptoms; the second stage occurs within days or weeks of the original infection and involves the skin and nervous, muscular, and skeletal systems; and the third stage occurs months or years after infection and may induce chronic arthritis, chronic neurologic disorders, and autoimmune disease. An asymptomatic infection has also been reported.

LYMErix was indicated for immunization against Lyme disease for individuals 15 to 70 years old. The vaccination consists of 3 intramuscular injections at 0, 1, and 12 months. The vaccination elicits antibodies against the outer surface lipoprotein A of *B. burgdorferi*. Since this epitope is not exposed on the bacterium once in the human body, it is thought that human antibodies reach the bacterium in the gut of the infected tick while it is biting an individual, thus preventing transmission of the bacterium.

In clinical trials LYMErix elicited immune protection against Lyme disease with an efficacy of 50% after two doses of the vaccine and 78% after three doses, and an efficacy of 100% against the asymptomatic infection after three doses. The most commonly reported side effects were pain and reactions at the injection site, flu-like symptoms, headaches, and muscle and joint pain. LYMErix should be administered only if necessary during pregnancy and lactation and is contraindicated in individuals receiving anticoagulant therapy.

Further Reading

http://www.fda.gov

http://www.gsk.com

Schoen, R.T. et al., Safety and immunogenicity profile of a recombinant outer-surface protein A Lyme disease vaccine: clinical trial of a 3-dose schedule at 0, 1, and 2 months, *Clin. Ther.*, 22, 315–325, 2000.

Sigal, L.H. et al., A vaccine consisting of recombinant Borrelia burgdorferi outer-surface protein A to prevent Lyme disease. Recombinant Outer-Surface Protein A Lyme Disease Vaccine Study Consortium, *N. Engl. J. Med.*, 339, 216–222, 1998.

Wallich, R. et al., The recombinant outer surface protein A (lipOspA) of Borrelia burgdorferi: a Lyme disease vaccine, *Infection*, 24, 396–397, 1996.

MabCampath/Campath-1H

Product Name: Alemtuzumab (international nonproprietary name)

MabCampath (trade name E.U.)

Campath-1H (trade name U.S.)

Description: MabCampath (Campath in the U.S.) is a humanized monoclonal antibody that binds the human CD52 antigen, a membrane protein expressed on T- and B-lymphocytes, monocytes, and macrophages. It is produced using recombinant DNA technology in Chinese hamster ovary (CHO) cells and supplied as a concentrated solution for intravenous infusion.

Approval Date: 2001 (E.U. and U.S.)

Therapeutic Indications: MabCampath is indicated for treatment of B-cell chronic lymphocytic leukemia (CLL) in patients who did not respond to alkylating agents and fludarabine phosphate therapy.

Manufacturer: Schering AG, Müllerstrasse 178, 13342 Berlin, Germany, http://www.schering.de (manufacturer is responsible for batch release in the European Economic Area)

Millennium and ILEX Partners, LP, 75 Sidney Street, Cambridge, MA 02139, http://www.millennium.com, http://www.ilexonc.com (U.S.)

Marketing: Millennium and ILEX UK Ltd., 1 & 3 Frederick Sanger Road, The Surrey Research Park, Guildford, Surrey GU2 7YD, U.K., http://www.millennium.com, http://www.ilexonc.com (E.U.)

Millennium and ILEX Partners, LP, 75 Sidney Street, Cambridge, MA 02139, http://www.millennium.com, http://www.ilexonc.com (U.S.)

Manufacturing

Alemtuzumab is a humanized monoclonal antibody. It was constructed by combining a human IgG1 kappa antibody with complementarity-determining regions (CDRs) from a rat monoclonal antibody specific for the human CD25 protein. The rat antibody was obtained as a result of genetic manipulation to enhance its binding activity for antigen. Alemtuzumab is produced in mammalian cell culture by using CHO cells. The purification process includes affinity, cation exchange and size exclusion chromatography, and viral inactivation and removal steps. The final product is presented as a concentrated solution for infusion containing alemtuzumab (active), disodium edetate (EDTA), polysorbate 80, and phosphate buffered saline as excipients.

The shelf life of the product is 36 months when stored at 2 to 8°C, protected from light. Routine evaluative tests are carried out on the final product to monitor appearance, purity, potency, sterility, presence of endotoxins, and viral and bacterial safety.

Overview of Therapeutic Properties

The most common form of leukemia affecting Western countries is CLL. It consists of proliferation and accumulation of malignant B cells in blood, bone marrow, and other tissues, leading to bone marrow dysfunction and enlargement of the lymph nodes, liver, and spleen. The indicated treatment for CLL is chemotherapy with alkylating agents (resulting in about 25% complete remission and 50% partial remission), and subsequently, in refractory patients, the use of purine analogues, mainly fludarabine. Patients who have no adequate response to treatment have a median survival of 6 to 9 months. Alemtuzumab is indicated for the treatment of patients who have had no satisfactory response to either therapy. The monoclonal antibody is specific for the CD52 antigen present on the surface of malignant and normal B lymphocytes, as well as T lymphocytes. CD52 is also present on monocytes, thymocytes, macrophages, and in a small percentage (fewer than 5%) of granulocytes (not on erythrocytes, platelets, or bone marrow stem cells). The cell lysis induced after antibody binding occurs only when alemtuzumab binds to lymphocytes.

Studies on CLL patients show a substantial reduction in the circulating lymphocyte mass and a significant effect in other tissues after alemtuzumab treatment. The response rate is around 30% but is higher at early stages of the disease. Alemtuzumab should be administered in escalating doses during the first week to reduce infusion-related reactions; thereafter, a 30-mg dose is administered every second day, 3 times a week, for up to 12 weeks.

The most common side effects were infusion-related reactions reported in more than 80% of the patients. Serious and, in rare instances, fatal hematological reactions have been reported, as well as serious and sometimes fatal bacterial and viral infections. Infections remain the highest cause of death in patients with CLL. No evidence of human antimouse antibody (HAMA) response was found. Alemtuzumab is contraindicated during pregnancy and lactation. Ongoing studies will investigate the intravenous vs. subcutaneous route of administration of alemtuzumab, as well as the efficacy of alemtuzumab vs. alkylating agents as front-line therapy in CLL patients.

Further Reading

http://www.campath.com
http://www.eudra.org
http://www.fda.gov
http://www.ilexonc.com
http://www.millennium.com

Dumont, F.J., Alemtuzumab (Millennium/ILEX), *Curr. Opin. Invest. Drugs*, 2, 139–160, 2001.

Flynn, J.M. and Byrd, J.C., Campath-1H monoclonal antibody therapy, *Curr. Opin. Oncol.*, 12, 574–581, 2000.

Osterborg, A. et al., Phase II multicenter study of human CD52 antibody in previously treated chronic lymphocytic leukemia. European Study Group of CAMPATH-1H Treatment in Chronic Lymphocytic Leukemia, *J. Clin. Oncol.*, 1567–1574, 1997.

MabThera/Rituxan

Product Name: MabThera (trade name E.U.)

Rituxan (trade name U.S.)

Rituximab (international nonproprietary name)

Description: MabThera (Rituxan in the U.S.) is a chimeric monoclonal antibody that binds to CD20, a transmembrane protein located on the surface of B lymphocytes. It is produced by recombinant DNA technology in mammalian cell culture. The glycosylated product consists of two 451-amino acid heavy chains and two 213-amino acid light chains and displays a molecular weight of 145 kDa. It is supplied as a 10-mg/ml concentrated solution for intravenous infusion.

Approval Date: 1997 (U.S.); 1998 (E.U.)

Therapeutic Indication: Rituximab is indicated for the treatment of patients with relapsed or refractory follicular non-Hodgkin's lymphoma who do not respond to conventional chemotherapy.

Manufacturer: Hoffmann-La Roche AG, Postfach 1270, 79630 Grenzach-Wyhlen, Germany (manufacturer is responsible for import and batch release in the European Economic Area) Genentech, Inc., 1 DNA Way, South San Francisco, CA 94080-4990, http://www.gene.com (manufacturer of finished product) and IDEC Pharmaceutical Corporation, 11011 Torreyana Road, San Diego, CA 92121, http://www.idecpharm.com (manufacturer of active substance) (U.S.)

Marketing: Roche Registration Limited, 40 Broadwater Road, Welwyn Garden City, Hertfordshire, AL7 3AY, U.K. (E.U.) Genentech, Inc., 1 DNA Way, South San Francisco, CA, http://www.gene.com and IDEC Pharmaceutical Corporation, 11011 Torreyana Road, San Diego, CA 92121, http://www.idecpharm.com (U.S.)

Manufacturing

The chimeric monoclonal antibody consists of variable regions from the 2H7 murine antibody and heavy and kappa constant domains from a human IgG1. The recombinant antibody is produced in mammalian cell culture using CHO cells. Control tests guarantee culture media safety; gentamycin is added against mycoplasma contamination, and transferrin (bovine-derived) has been replaced by the use of a higher amount of ferrous sulphate in the culture media. The purification process includes protein A affinity chromatography, anion exchange chromatography, and viral removal procedures. The final product is a concentrated solution for infusion containing rituximab (active), as well a the following excipients: sodium citrate, polysorbate 80, sodium chloride, and sodium hydroxide and hydrochloric acid to adjust the pH to 6.5.

The shelf life of the product is 24 months when stored at 2 to 8°C, protected from direct sunlight. Routine evaluative tests carried out during processing and on the final product include amino acid analysis, peptide mapping, analysis of oligosaccharides, cIDF, SDS-PAGE, CD, tests for the presence of protein A, bovine IgG, endotoxins, bioburden, and viral and bacterial safety.

Overview of Therapeutic Properties

Patients with non-Hodgkin's lymphoma, particularly the follicular lymphoma, often exhibit several relapses and ultimately develop a chemoresistant lymphoma. Rituximab, with its novel mechanism of action, is indicated in patients who exhibited no adequate response to conventional treatments, including chemotherapy, drugs like interferon-alfa, and autologous bone marrow transplantation. The recombinant antibody mediates cell lysis by selectively binding the CD20 antigen on the surface of normal and malignant B cells (except myeloma cells and proB cells). The chimeric antibody has the advantage over the murine antibody of reduced immunogenicity, reduced toxicity (low efficacy of the murine antibody required radiolabelling), and increased efficacy, with the human domains activating human complement and mediating cytotoxicity.

The short duration of administration of rituximab, once a week for 4 weeks, is an advantage compared to the 4 to 6 months required for conventional treatment. Studies in patients with non-Hodgkin's lymphoma show that a rapid and selective depletion of B cells is observed within the first day post-infusion, with a response rate of 57% in patients with follicular lymphoma and 11% in patients with small B-lymphocytic lymphoma.

Infusion-related symptoms consisting of a cytokine release syndrome with fever, chills, shivering, flushing, etc., which were usually reversible after cessation of the treatment with Rituximab, were observed in more than 50% of patients. Of those patients, 10% showed hypotension and bronchospasm. A severe cytokine release syndrome was reported in patients with high numbers of circulating tumor cells. An increased incidence of mild infections was observed after treatment, but fewer than after conventional chemotherapy. Rarely, cardiac failure, breathing difficulties, severe skin reactions, and acute renal failure, observed after infusion of rituximab, had fatal outcomes. Rituximab is contraindicated during pregnancy.

Further Reading

http://www.eudra.org
http://www.fda.gov
http://www.gene.com
http://www.rituxan.com
Grillo-Lopez, A.J. et al., Rituximab: the first monoclonal antibody approved for the treatment of lymphoma, *Curr. Pharm. Biotechnol.*, 1, 1–9, 2000.
Maloney, D.G. et al., IDEC-C2B8: results of a phase I multiple-dose trial in patients with relapsed non-Hodgkin's lymphoma, *J. Clin. Oncol.*, 15, 3266–3274, 1997.

Metalyse/TNKase

Product Name: Metalyse (trade name E.U.)

TNKase (trade name U.S.)

Tenecteplase (international nonproprietary name)

Description: Metalyse (TNKase in the U.S.) is a recombinant 525-amino acid glycoprotein derived from the human tissue plasminogen activator (tPA). It differs from the natural molecule in the residues at position 103 (a threonine residue replaced by an asparagine), 117 (asparagine replaced by glutamine), 296 (lysine replaced by alanine), 297 (histidine replaced by alanine), and 298 and 299 (arginines replaced by alanines). It is produced by recombinant DNA technology in Chinese hamster ovary (CHO) cells. The final product is presented in a lyophilized form to be reconstituted using the provided solvent-filled syringe before intravenous administration.

Approval Date: 2000 (U.S.); 2001 (E.U.)

Therapeutic Indications: Metalyse is indicated in the thrombolytic therapy of acute myocardial infarction.

Manufacturer: Boehringer Ingelheim Pharma KG, Birkendorferstrasse 65, 88397 Biberach/Riss, Germany, http://www.boehringer-ingelheim.com (manufacturer is responsible for batch release in the European Economic Area)

Genentech, Inc., 1 DNA Way, South San Francisco, CA 94080-4990, http://www.gene.com (U.S.)

Marketing: Boehringer Ingelheim International GmbH, Binger Strasse 173, 55216, Ingelheim am Rhein, Germany, http://www.boehringer-ingelheim.com (E.U.)

Genentech, Inc., 1 DNA Way, South San Francisco, CA 94080-4990, http://www.gene.com (U.S.)

Manufacturing

Metalyse is a recombinant form of the human tPA. It differs from the natural molecule in the substitutions of the amino acid residues at positions 103 (Thr to Asn), 117 (Asn to Gln), 296 (Lys to Ala), 297 (His to Ala), and 298 and 299 (Arg to Ala). The recombinant product is expressed in a CHO cell line, and its purification involves several chromatographic steps, ultrafiltrations, and procedures to inactivate and remove viral contaminants. The final product contains the active substance tenecteplase, as well as L-arginine, phosphoric acid, and polysorbate 80 as excipients. Metalyse is provided in a lyophilized form to be reconstituted using the prefilled sterile syringe for intravenous administration.

The shelf life of the product is 24 months when stored at temperatures lower than 30°C. Extensive tests have been carried out to remove viral contaminants and to ensure the quality of the final product.

Overview of Therapeutic Properties

Metalyse is a thrombolytic agent administered in cases of acute myocardial infarction. It is administered within 6 hours of appearance of symptoms, and it catalyses the degradation of plasminogen to plasmin, a protease that degrades the fibrin clot.

Metalyse has been modified to exhibit increased affinity for fibrin relative to the natural molecule. This increases its half-life and enables a single bolus administration. Furthermore, it is more resistant to plasminogen activator inhibitors and therefore has an increased efficacy over the natural molecule.

Metalyse should be administered intravenously as a single bolus at doses suited to the individual needs of patient according to body weight. Acetylsalicylic acid and heparin should be given after the administration of metalyse.

Metalyse proved to be as effective as another thrombolytic agent, alteplase, in reducing the mortality of acute miocardial infarction, with the advantage of a decreased incidence of nonintracranial bleedings. Bleedings, mainly at the injection site, and arrhythmias remained the most common side effects, as they are with other thrombolytic agents. Metalyse should not be administered during pregnancy and lactation.

Further Reading

http://www.boehringer-ingelheim.com
http://www.eudra.org
http://www.fda.gov

http://www.gene.com
http://www.tnkase.com
Angeia, B.G. et al., Safety of the weight-adjusted dosing regimen of tenecteplase in the ASSENT-Trial, *Am. J. Cardiol.*, 88, 1240–1245, 2001.
Davydov, L. and Cheng J.W., Tenecteplase: a review, *Clin. Ther.*, 23, 982–997, 2001.
Van de Werf, F. et al., Incidence and predictors of bleeding events after fibrinolytic therapy with fibrin-specific agents: a comparison of TNK-tPA and rt-PA, *Eur. Heart J.*, 22, 2253–2261, 2001.

Mylotarg

Product Name:	Mylotarg (trade name) Gemtuzumab ozogamicin (international nonproprietary name)
Description:	Mylotarg is a 153-kDa immunotoxin consisting of a humanized monoclonal antibody (hP67.6) linked to a bacterial toxin, calicheamicin, with cytotoxic effect. The humanized antibody specifically binds the human CD33 antigen, which is expressed on normal and leukemic myeloid cells, and is produced in recombinant form in a modified murine NSO cell line. Mylotarg is presented as a lyophilized powder (5 mg/vial) to be resuspended before intravenous infusion.
Approval Date:	2000 (U.S.)
Therapeutic Indications:	Mylotarg is indicated for the treatment of acute myeloid leukemia in CD33-positive patients who are over 60 years old at the time of their first relapse and are not candidates for standard chemotherapy.
Manufacturer:	Wyeth-Ayerst Laboratories, P.O. Box 8299, Philadelphia, PA, http://www.wyeth.com
Marketing:	Wyeth-Ayerst Laboratories, P.O. Box 8299, Philadelphia, PA, http://www.wyeth.com

Manufacturing

The immunotoxin gemtuzumab ozogamicin consists of a recombinant humanized monoclonal antibody linked to the cytotoxin calicheamicin. The antibody (hP67.6) is produced in a modified murine NSO cell line using recombinant DNA technology and consists of a human IgG4 antibody with complementarity-determining regions (CDRs) from a murine monoclonal antibody (p67.6) specific for the human CD33 antigen. The bacterial toxin calicheamicin is obtained from a *Micromonospora echinospora* culture. After purification, which includes chromatographic and ultrafiltration procedures, the antibody and toxin are chemically linked via a bifunctional linker group. The final product contains gemtuzumab ozogamicin (active), as well as dextran 40, sucrose, sodium chloride, monobasic sodium phosphate, and dibasic sodium phosphate as excipients.

Extensive tests are carried out during processing and on the final product to check purity, potency, sterility, presence of endotoxins, and for viral and bacterial safety.

Overview of Therapeutic Properties

Acute myeloid leukemia is a common form of leukemia in adults. It consists of an accumulation of abnormal white blood cells in the blood and bone marrow, with a high incidence of relapses and fatalities. Gemtuzumab ozogamicin is the first immunotoxin approved in the U.S. that consists of an antibody linked to a bacterial toxin. The antibody is specific for the antigen CD33, which is expressed on the surface of cancerous and normal myeloid cells in 80% of acute myeloid leukemia patients. Following antibody binding, the immunotoxin is internalized into the myeloid cells, at which point the potent toxin calicheamicin binds to DNA and breaks the double strand, thus inducing cell death. Mylotarg is indicated in the treatment of acute myeloid leukemia in CD33-positive patients who are over 60, undergoing their first relapse, and who are not considered candidates for standard chemotherapy.

Mylotarg should be administered twice as 2-hour intravenous infusions, with an interval of 14 days between the doses. Though clinical studies did not involve a direct comparison between Mylotarg treatment and standard chemotherapy, Mylotarg exhibited response and survival rates similar to chemotherapy, but necessitated fewer administrations and led to the elimination of side effects associated with standard chemotherapy. Clinical studies showed that administration of Mylotarg resulted in an overall response of 30%, including 26% in 60-year-old patients, with an overall median duration of survival of 5.9 months.

The most commonly reported side effects were infusion-related reactions. The use of Mylotarg induces myelosuppression due to the target of normal as well as cancerous myeloid cells. Myelosuppression is temporary and reversible, since pluripotent hemapoietic stem cells are not targeted by Mylotarg, and values revert to normal at the end of treatment. A prophylactic treatment is recommended to avoid the severe effects of myelosuppression (anemia, low platelet counts, infection, bleeding), as well as hematologic monitoring. Mylotarg is contraindicated during pregnancy and lactation.

Further Reading

http://www.fda.gov
http://www.wyeth.com
Lang, K. et al., Outcomes in patients treated with gemtuzumab ozogamicin for relapsed acute myelogenous leukemia, *Am. J. Health Syst. Pharm.*, 59, 941–948, 2002.
Larson, R.A. et al., Antibody-targeted chemotherapy of older patients with acute myeloid leukemia in first relapse using Mylotarg (gemtuzumab ozogamicin), *Leukemia*, 16, 1627–1636, 2002.
Sievers, E.L. et al., Efficacy and safety of gemtuzumab ozogamicin in patients with CD33-positive acute myeloid leukemia in first relapse, *J. Clin. Oncol.*, 19, 3244–3254, 2001.
Stadtmauer, E.A., Gemtuzumab ozogamicin in the treatment of acute myeloid leukemia, *Curr. Oncol. Rep.*, 4, 375–380, 2002.

Myoscint (withdrawn from market)

Product Name:	Myoscint (trade name) Imciromab pentetate (international nonproprietary name)
Description:	Myoscint consists of an antigen-binding Fab fragment derived from a murine IgG2 kappa monoclonal antibody raised against human myosin. The antibody fragment is conjugated to a linker-chelator for coupling to the radioisotope Indium 111 (In 111) before use. The monoclonal antibody is produced in a hybridoma cell line. Myoscint is provided as a solution for intravenous administration for radioimaging purposes in a kit for reconstitution of the radiolabeled molecule. The radioisotope is not provided.
Approval Date:	1996 (U.S.)
Withdrawal Date:	1999 (for commercial reasons)
Therapeutic Indications:	Myoscint is coupled to Indium 111 for use as a cardiac agent to detect the presence and location of myocardial injury in patients with suspected myocardial infarction.
Manufacturer:	Centocor B.V., Einsteinweg 101, 2333 CB Leiden, the Netherlands, http://www.centocor.com
Marketing:	Centocor, Inc., 200 Great Valley Parkway, Malvern, PA 19355, http://www.centocor.com

Manufacturing

Myoscint consists of a Fab fragment of a monoclonal antibody against human cardiac myosin, conjugated to a linker-chelator for coupling with the radioisotope Indium 111 (In 111). The murine monoclonal antibody is produced by standard hybridoma technology using B cells taken from the spleen of mice immunized with human cardiac myosin fused to a myeloma cell line. The purified antibody undergoes enzymatic digestion with papain to generate Fab fragments, which are then conjugated to the linker-chelator, diethylenetriaminepentaacetic acid. Purification involves several chromatographic steps and other procedures to ensure viral safety. The final product is presented in a kit containing a solution of imciromab pentetate (active), as well as a sodium phosphate buffer, sodium chloride, and maltose as excipients, and a second solution containing a sodium citrate buffer. The latter is mixed with the antibody solution before coupling with the radioisotope In 111, which is not supplied.

Myoscint has a shelf life of 18 months when stored at 2 to 8°C. Extensive tests were carried out to verify the quality and safety of the product, including mass spectroscopy, amino acid analysis, tryptic mapping, monosaccharide analysis, circular dichroism, SDS-PAGE, GF-HPLC, IEF, and other tests to ensure its microbiological and viral safety.

Overview of Therapeutic Properties

Myoscint was indicated for diagnostic radioimaging to detect injured heart muscle in patients with suspected myocardial infarction. Cardiac myosin is found exclusively in the intracellular compartment of myocardial cells; therefore, the specificity of the antibody for human cardiac myosin allows detection of dead heart tissue. The use of the smaller Fab fragment of the antibody allows greater tissue penetration and more rapid clearance from the circulation than the whole antibody. Radiolabeled Myoscint should be administered as an intravenous bolus, and radioimaging is performed 18 to 24 hours after administration, or after 48 hours when insufficient blood clearance has occurred after 24 hours.

The efficiency of Myoscint was demonstrated in detecting and localizing injured myocardial and skeletal tissue with great specificity. Myoscint proved to be safe and well tolerated. Pain at the site of injection, chest pain, headaches, and fever were the most commonly reported side effects. No human antimurine antibodies have been detected after administration of Myoscint. Myoscint should be used only if necessary in pregnant women, and its use is contraindicated during lactation.

Further Reading

http://www.centocor.com
http://www.fda.gov
Morguet, A.J. et al., Immunoscintigraphy with 111In antimyosin Fab, *Nucl. Med. Commun.*, 11, 727–735, 1990.

Taillefer, R., Detection of myocardial necrosis and inflammation by nuclear cardiac imaging, *Cardiol. Clin.*, 12, 289–30, 1994.

Yamada, T. et al., Time course of myocardial infarction evaluated by indium-111-antimyosin monoclonal antibody scintigraphy: clinical implications and prognostic value, *J. Nucl. Med.*, 33, 1501–1508, 1992.

Natrecor

Product Name: Natrecor (trade name)

Nesiritide (common name)

Description: Natrecor is a recombinant form of human B-type natriuretic peptide (hBNP), a 32-amino acid, 3.464-kDa peptide produced naturally by the ventricular myocardium. Natrecor displays an identical amino acid sequence to the native molecule. The peptide promotes the dilation of veins and arteries and is used to alleviate shortness of breath associated with congestive heart failure. Natrecor is presented as a lyophilized powder in single-use vials, each containing 1.5 mg active ingredient.

Approval Date: 2001 (U.S.)

Therapeutic Indications: Natrecor is indicated for the treatment of patients with acutely decompensated congestive heart failure who have dyspnea (labored or difficult breathing) at rest or with minimal activity. In such patients, intravenous administration of Natrecor reduced pulmonary capillary wedge pressure and improved dyspnea.

Manufacturer: Scios Inc., 820 West Maude Avenue, Sunnyvale, CA 94085, http://www.sciosinc. com/home

Marketing: Scios Inc., 820 West Maude Avenue, Sunnyvale, CA 94085, http://www.sciosinc. com/home

Manufacturing

A nucleotide sequence coding for the 32 amino acid hBNP is expressed in a recombinant *Escherichia coli* strain. After initial fermentation and product extraction, the peptide is purified using a combination of high resolution chromatographic steps. Mannitol, citric acid, and sodium citrate are added as excipients. The finished product is filter sterilized, aseptically transferred into presterile vials, and lyophilized.

Overview of Therapeutic Properties

Natrecor (hBNP) has proved effective in treating dyspnea associated with conjestive heart failure. The peptide brings about its effect by binding to the particulate guanylate cyclase receptor found on the surface of smooth muscle and endothelial cells. Binding results in the generation of increased intracellular concentrations of 3′ – 5′ adenylate cyclic GMP (cGMP), which in turn promotes smooth muscle cell relaxation. This results in dilation of veins and arteries. Upon i.v. administration, hBNP displays a mean terminal elimination half-life ($t_{1/2}$) of approximately 18 minutes. It is eliminated from the circulation via three different mechanisms: binding to cell surface clearance receptors, with subsequent internalization and lysosomal proteolytic degradation; proteolysis by endopeptidases found on the vascular lumen surface; or renal clearance.

The effects of Natrecor have been studied in 10 clinical trials. The product was generally found to reduce pulmonary capillary wedge syndrome and to improve dyspnia. Prior to administration, the content of a single-product vial is reconstituted in a final volume of 250 ml of a diluent of choice, such as 5% dextrose solution. Natrecor is generally administered as an initial i.v. bolus dose of 2 μg/kg body weight, followed by continuous infusion of the remainder at a rate of 0.01 μg/kg/min.

The product is contraindicated in patients with a known hypersensitivity to any of its components. The safety and effectiveness of Natrecor has not been assessed in pediatric patients, and it should be administered to nursing mothers with caution. Adverse reactions, usually occurring in 3% of patients, can include hypotension, ventricular tachycardia, and angina pectoris. It may also affect renal function and induce abdominal and back pain, as well as dizziness, nausea, and a state of anxiety.

Further Reading

Anon., Natrecor — Nesiritide for injection SCIOS — recombinant natriuretic peptide for tx of acutely decompensated CHF, *Formulary*, 36(9), 628–628, 2001.

Keating, G.M. and Goa, K.L., Nesirifide — a review of its use in acute decompensated heart failure, *Drugs*, 63(1), 47–70, 2003.

Mills, R.M. and Hobbs, R.E., How to use nesiritide in treating decompensated heart failure, *Clev. Clin. J. Med.*, 69(3), 252–256, 2002.

Vichiendilokkul, A. et al., Nesiritide: a novel approach for acute heart failure, *Ann. Pharmacother.*, 37(2), 247–258, 2003.

NeoRecormon

Product Name: NeoRecormon (trade name)

Epoetin beta (international nonproprietary name)

Description: NeoRecormon is a recombinant human epoetin beta (erythropoietin) that stimulates the production of red blood cells. NeoRecormon is identical to the natural molecule, as derived from the urine of anemic patients. NeoRecormon is produced in Chinese hamster ovary (CHO) cells, and it is presented either in a lyophilized form or as a solution for intravenous or subcutaneous administration. NeoRecormon replaced Recormon, which contained the same active substance. NeoRecormon is offered in different formulations, including a prefilled, ready-to-use syringe and devices for multidose injections.

Approval Date: 1997 (E.U.)

Therapeutic Indications: NeoRecormon is indicated for the treatment of renal anemia in patients undergoing dialysis or prior to initiation of dialysis. It is also indicated for prevention and treatment of anemia in premature infants and in adults with solid tumors, multiple myeloma, non-Hodgkin's lymphoma, and chronic lymphocytic leukemia who are undergoing antitumor therapy, as well as for autologous donation.

Manufacturer: Roche Diagnostics GmbH, Sandhofer Strasse 116, 68305 Mannheim, Germany, http://www.roche-diagnostics.com (manufacturer is responsible for batch release in the European Economic Area)

Marketing: Roche Registration Limited, 40 Broadwater Road, Welwyn Garden City, Hertfordshire AL7 3AY, U.K., http://www.roche.com

Manufacturing

NeoRecormon is a recombinant human epoetin beta, identical to the natural molecule. It is produced in a modified CHO cell line. As the protein is expressed in mammalian cells, it retains the glycosylation pattern and the characteristics and activity of the natural protein. Purification of the expressed protein involves several chromatographic steps, including Blue Sepharose dye affinity chromatography, Hydroxyapatite, RP-HPLC, and DEAE Sepharose chromatography. The final product contains the active substance epoetin beta, as well as sodium chloride, calcium chloride, sodium dihydrogen phosphate, sodium monohydrogen phosphate, glycine, leucine, isoleucine, threonine, glutamic acid, phelylalanine, urea, and polysorbate 20 as excipients. The active substance is presented in a number of forms: lyophilized with a solvent provided for reconstitution, as a solution ready for injection in prefilled sterile syringes, and in cartridges to be used with a pen for multidose injection.

The shelf life of the product is 36 months when stored at 2 to 5°C. The product has been extensively tested to ensure quality and the absence of viral and microbial contaminants.

Overview of Therapeutic Properties

NeoRecormon contains epoetin beta, a recombinant human hormone stimulating the production of red blood cells. It is indicated in the treatment of renal anemia and for the prevention and treatment of anemia in premature infants, in patients undergoing antitumor therapy, and for autologous blood donation.

NeoRecormon should be administered weekly as an intravenous or subcutaneous injection for the length of time required according to clinical indications. NeoRecormon proved its efficacy in the treatment and prevention of anemia, with increased hematocrit values observed. The treatment regimen also includes administration of iron. NeoRecormon was determined to be safe, and the most common side effects were increased blood pressure, infections in the respiratory tract, and increased platelet counts.

Further Reading

http://www.eudra.org
http://www.roche.com

Dunn, C.J. and Markham, A., Epoetin beta. A review of its pharmacological properties and clinical use in the management of anaemia associated with chronic renal failure, *Drugs*, 51, 299–318, 1996.

Locatelli, F. et al., Once-weekly compared with three-times-weekly subcutaneous epoetin beta: results from a randomized, multicenter, therapeutic-equivalence study, *Am. J. Kidney Dis.*, 40, 119–125, 2002.

Maier, R.F. et al., The effect of epoetin beta (recombinant human erythropoietin) on the need for transfusion in very-low-birth-weight infants. European Multicentre Erythropoietin Study Group, *N. Engl. J. Med.*, 330, 1173–1178, 1994.

Osterborg, A. et al., Recombinant human erythropoietin in transfusion-dependent anemic patients with multiple myeloma and non-Hodgkin's lymphoma — a randomized multicenter study. The European Study Group of Erythropoietin (Epoetin Beta) Treatment in Multiple Myeloma and Non-Hodgkin's Lymphoma, *Blood*, 87, 2675–2682, 1996.

Osterborg, A. et al., Randomized, double-blind, placebo-controlled trial of recombinant human erythropoietin, epoetin Beta, in hematologic malignancies, *J. Clin. Oncol.*, 20, 2486–2494, 2002.

Nespo/Aranesp

Product Name: Nespo (trade name E.U.)

Aranesp (trade name U.S.)

Darbepoetin alfa (international nonproprietary name)

Description: Nespo (Aranesp in the U.S.) is a recombinant 166-amino acid human erythropoietin that stimulates the production of red blood cells. It differs from the natural molecule in the occurrence of two additional sugar chains. It has five N-linked carbohydrate side chains, while the natural molecule has only three. This leads to an increased half-life and, therefore, a coupled, reduced frequency of administration is necessary. Nespo is produced in Chinese hamster ovary (CHO) cells and is presented as a solution for intravenous or subcutaneous administration.

Approval Date: 2001 (E.U. and U.S.)

Therapeutic Indications: Nespo is indicated for the treatment of anemia in patients with chronic renal failure, and now (in the U.S.) for patients with anemia associated with malignancies.

Manufacturer: Amgen Europe B.V., Minervum 7061, 4817 ZK Breda, the Netherlands, http://www.amgen.com (manufacturer is responsible for import and batch release in the European Economic Area)
Amgen, Inc., One Amgen Center Drive, Thousand Oaks, CA, amgen.com (U.S.)

Marketing: Dompé Biotec S.p.A., Via Santa Lucia 4, 20122 Milan, Italy, http://www.dompe.com (E.U.)
Amgen, Inc., One Amgen Center Drive, Thousand Oaks, CA, amgen.com (U.S.)

Manufacturing

Nespo is a recombinant human erythropoietin (rHuEPO) produced by recombinant DNA technology in a modified CHO cell line. It differs from the natural protein in its increased carbohydrate content. The recombinant protein is collected from the culture medium and is purified using several chromatographic steps, as well as a number of procedures to inactivate and remove viral and microbial contaminants. The final product contains the active substance darbepoetin alfa and the following excipients: monobasic sodium phosphate, dibasic sodium phosphate, sodium chloride, and polysorbate 80. The product is provided as a solution in a vial or in a sterile prefilled syringe for intravenous or subcutaneous administration.

The shelf life of the product is 24 months when stored at 2 to 8°C. The quality and the safety of the product have been ensured by extensive control tests.

Overview of Therapeutic Properties

Nespo is a recombinant erythropoietic hormone that stimulates the production of red blood cells. It is indicated in the treatment of anemia in adults and children with chronic renal failure and, in the U.S. only, for patients suffering from anemia due to malignancies or antitumor treatment. Nespo has been modified in order to increase its carbohydrate content. These modifications increase its half-life *in vivo*, which allows for less frequent administration of the drug than is necessary in the case of the natural or the rHuEPO molecules.

Nespo should generally be administered intravenously or subcutaneously once a week, but in some cases once every two weeks is sufficient. The treatment consists of a correction phase, during which the necessary dose should be adjusted to the individual needs of the patient in order to increase the hemoglobin levels, and a maintenance phase to maintain the elevated hemoglobin values in the longer term. Iron supplements may be required, and the deficiency of folic acid or vitamin B12 should be corrected in order not to compromise the efficiency of Nespo.

Nespo, with the same specificity and activity of rHuEPO, proved its efficacy in increasing and maintaining the levels of hemoglobin to the same extent as the natural molecule. The most common side effects were the same as those observed after administration of other erythropoietic products and include hypertension and vascular access thrombosis. Pain was also reported at the site of injection when Nespo was administered subcutaneously. Care should be taken when Nespo is administered to patients with liver failure and during pregnancy. Nespo is contraindicated in patients with blood pressure disorders and during lactation.

Further Reading

http://www.amgen.com
http://www.aranesp.com
http://www.eudra.org
http://www.fda.gov
Glaspy, J.A. et al., Darbepoetin alfa given every 1 or 2 weeks alleviates anaemia associated with cancer chemotherapy, *Br. J. Cancer*, 87, 268–276, 2002.
Joy, M.S., Darbepoetin alfa: a new erythropoiesis-stimulating protein, *Ann. Pharmacother.*, 36, 1183–1192, 2002.
Nissenson, A.R. et al., Ransomized, controlled trial of darbepoetin alfa for the treatment of anemia in hemodialysis patients, *Am. J. Kidney Dis.*, 40, 110–118, 2002.

Neulasta

Product Name: Neulasta (trade name)

Pegfilgrastim (international nonproprietary name)

Description: Neulasta is a recombinant human granulocyte-colony stimulating factor (G-CSF), conjugated to a single 20-kDa methoxypolyethylene glycol-propionaldehyde (PEG) molecule at the N terminus. A nonpegylated form of the molecule is marketed in the U.S. under the trade name Neupogen. The 175-amino acid G-CSF is produced by recombinant DNA technology in *Escherichia coli*, followed by *in vitro* attachment of the PEG molecule. Like Neupogen, Neulasta stimulates the production of neutrophils in the blood. Neulasta is presented in prefilled syringes as a solution for subcutaneous injections.

Approval Date: 2002 (E.U. and U.S.)

Therapeutic Indications: Neulasta is indicated for the treatment of neutropenia and febrile neutropenia in patients with malignancies treated with cytotoxic chemotherapy.

Manufacturer: Amgen Europe B.V., Minervum 7061, 4817 ZK Breda, the Netherlands, http://www.amgen.com (manufacturer is responsible for import and batch release in the European Economic Area)
Amgen, Inc., One Amgen Center Drive, Thousand Oaks, CA, http://www.amgen.com (U.S.)

Marketing: Amgen Europe B.V., Minervum 7061, 4817 ZK Breda, the Netherlands, http://www.amgen.com (E.U.)
Amgen, Inc., One Amgen Center Drive, Thousand Oaks, CA, http://www.amgen.com (U.S.)

Manufacturing

Pegfilgrastim is a Pegylated form of human G-CSF produced by recombinant DNA technology in *E. coli*. G-CSF is a nonglycosylated molecule and contains an additional methionine residue at the N terminus of the natural human amino acid sequence. Pegfilgrastim is obtained by coupling a single 20-kDa PEG molecule to the N terminus of filgrastim, following which the Pegylated molecule is purified by cation exchange chromatography and extensively characterized.

The final product is presented in prefilled syringes as a solution for subcutaneous injection and contains the active substance Pegfilgrastim, as well as sodium acetate, sorbitol, and polysorbate 20. The shelf life of the product is 24 months when stored at 2 to 8°C, protected from light. The quality and the safety of the product are ensured by extensive tests.

Overview of Therapeutic Properties

Neulasta is indicated in the treatment of neutropenia and febrile neutropenia in adult patients with malignancy who are receiving anticancer cytotoxic chemotherapy. Its active substance, pegfilgrastim, a sustained duration form of G-CSF (filgrastim) as Pegylation, reduced its rate of renal clearance. Like filgrastim, it stimulates the production of white blood cells in the bone marrow. Neulasta should be administered as a subcutaneous injection 24 hours after the last dose of chemotherapy, after each cycle of chemotherapy. The reduced renal clearance of Pegfilgrastim, compared to filgrastim, results in a prolonged action. Therefore, a single dose injected at the end of each cycle of chemotherapy suffices with pegfilgrastim, while daily injections for up to 14 days are necessary in the case of filgrastim. Pegfilgrastim proved to be as effective as filgrastim in increasing peripheral blood neutrophil cell counts within 24 hours of administration.

Bone pain was the most common side effect related to administration of Neulasta, followed by pain at the site of injection and general pain and headache. These occur with the same frequencies reported for filgrastim. No neutralizing antibodies were detected after administration of Neulasta. Neulasta should not be administered during pregnancy and lactation and to patients less than 18 years old.

Further Reading

http://www.amgen.com
http://www.eudra.org

http://www.fda.gov
http://www.neulasta.com
Bence, A.K. and Adams, V.R., Pegfilgrastim: a new therapy to prevent neutropenic fever, *J. Am. Pharm. Assoc. (Wash.)*, 42, 806–808, 2002.
Curran, M.P. and Goa, K.L., Pegfilgrastim, *Drugs*, 62, 1207–1215, 2002.

Neumega

Product Name:	Neumega (trade name) Oprelvekin (international nonproprietary name)
Description:	Neumega is a recombinant form of human interleukin 11 (IL-11) that lacks the N-terminal amino acid (proline) and glycocomponent of the natural molecule. The 19-kDa, 177-amino acid polypeptide is produced in *Escherichia coli* using recombinant DNA technology. Neumega is provided as a lyophilized powder (5 mg/vial) to be resuspended before subcutaneous injection.
Approval Date:	1997 (U.S.)
Therapeutic Indications:	Neumega is indicated for the prevention of severe thrombocytopenia and the reduction of platelet transfusions in patients with nonmyeloid malignancies after myelosuppressive chemotherapy.
Manufacturer:	Genetics Institute, Inc., One Burtt Road, Andover, MA, http://www.genetics.com (U.S.)
Marketing:	Genetics Institute, Inc., One Burtt Road, Andover, MA, http://www.genetics.com (U.S.)

Manufacturing

Neumega, a modified form of human IL-11, is produced in *E. coli* using recombinant DNA technology as a thioredoxin/IL-11 fusion protein. The fusion protein is then cleaved and IL-11 is purified using a number of chromatographic steps. The final product is provided as a lyophilized powder containing oprelvekin (active) as well as glycine, heptahydrate dibasic sodium phosphate, and monohydrate monobasic sodium phosphate as excipients. Neumega is reconstituted before subcutaneous administration.

Extensive tests are conducted to ensure the quality and the safety of the product.

Overview of Therapeutic Properties

Neumega is a recombinant form of human IL-11. It is very similar to the naturally occurring molecule and retains its activity in stimulation of the proliferation of hematopoietic stem cells and megakaryocyte progenitor cells and induction of the maturation of megakaryocytes, leading to an increase in platelet production. Neumega is used to increase the platelet count in the blood and to reduce the need for platelet transfusions after myelosuppressive chemotherapy treatment in patients with nonmyeloid malignancies. It is administered subcutaneously daily for up to 21 days, 6 to 24 hours after the completion of each cycle of chemotherapy. Neumega was found to be effective in the treatment of severe thrombocytopenia and in reducing the need for platelet transfusions, but without affecting disease progression or survival.

The most commonly reported side effects were fluid retention, which may have serious consequences in patients with congestive heart failure, and abnormal heart rhythms. These side effects disappeared after the treatment ended. Neumega is not indicated for pediatric use and is contraindicated during pregnancy and lactation.

Further Reading

http://www.fda.gov
http://www.genetics.com
http://www.neumega.com

Isaacs, C. et al., Randomized placebo-controlled study of recombinant human interleukin-11 to prevent chemotherapy-induced thrombocytopenia in patients with breast cancer receiving dose-intensive cyclophosphamide and doxorubicin, *J. Clin. Oncol.*, 15, 3368–3377, 1997.

Kaye, J.A., FDA licensure of NEUMEGA to prevent severe chemotherapy-induced thrombocytopenia, *Stem Cells*, 16(Suppl. 2), 207–223, 1998.

Sitaraman, S.V. and Gewirtz, A.T., Oprelvekin. Genetics Institute, *Curr. Opin. Investig. Drugs*, 2, 1395–1400, 2001.

Neupogen

Product Name:	Neupogen (trade name) Filgrastim (international nonproprietary name)
Description:	Filgrastim is a recombinant human granulocyte-colony stimulating factor (G-CSF) that regulates the production of neutrophils. The molecule is produced in *Escherichia coli* using recombinant DNA technology and differs from the naturally occurring molecule in having an additional methionine residue at the N-terminus and in lacking glycosylation. Neupogen is provided as a solution for subcutaneous or intravenous administration in vials and in prefilled syringes.
Approval Date:	1991 (U.S.)
Therapeutic Indications:	Neupogen is indicated for the treatment of neutropenia associated with various medical conditions.
Manufacturer:	Amgen, Inc., One Amgen Center Drive, Thousand Oaks, CA, http://www.amgen.com (U.S.)
Marketing:	Amgen, Inc., One Amgen Center Drive, Thousand Oaks, CA, http://www.amgen.com (U.S.)

Manufacturing

Filgrastim is a recombinant human G-CSF produced in *E. coli* using recombinant DNA technology. *E. coli* cells producing filgrastim are harvested and lysed, following which filgrastim is purified using several chromatographic and ultrafiltration procedures. The final product is provided as a solution for subcutaneous or intravenous administration and contains filgrastim (active), as well as mannitol, polysorbate 80, and sodium acetate as excipients.

The shelf life of the product is 24 months when stored at 2 to 8°C. Extensive testing ensures the quality and the safety of the product, including SDS-PAGE, HPLC, amino acid analysis, N-terminal sequence analysis, peptide mapping, and Western blotting.

Overview of Therapeutic Properties

The recombinant molecule filgrastim is very similar to natural human G-CSF and exhibits the same activity, consisting of stimulating the production of neutrophils from the bone marrow. Filgrastim is used in the case of febrile neutropenia in patients with nonmyeloid malignancies who receive myelosuppressive chemotherapy. It is administered daily, either subcutaneously or intravenously, 24 hours after each cycle of chemotherapy for up to 14 days. Filgrastim was found to reduce fever, infection, use of antibiotics, and hospitalization without affecting the progression of the disease or survival.

Filgrastim was found to reduce the median number of days of severe neutropenia observed in patients with severe chronic neutropenia, acute myeloid leukemia, and nonmyeloid malignancies who receive myeloablative chemotherapy followed by bone marrow transplant. Patients who received filgrastim before engraftment of mobilized peripheral blood progenitor cells had reduced lengths of platelet transfusions and red blood cell transfusions and shorter duration of post-transplant hospitalization.

The most commonly reported side effects were bone and muscular pain. Elevation of serum uric acid was also commonly reported. Antibodies against filgrastim have only rarely been reported. Rare cases of splenic rupture have been reported in patients who received filgrastim for peripheral blood progenitor cell mobilization. Neupogen is contraindicated during pregnancy and lactation.

Further Reading

http://www.amgen.com

http://www.fda.gov

http://www.neupogen.com

Dale, D.C., Colony-stimulating factors for the management of neutropenia in cancer patients, *Drugs*, 62(Suppl. 1), 1–15, 2002.

Foote, M. and Boone, T., Biopharmaceutical drug development: a case history, in *Biopharmaceuticals, an Industrial Perspective*, Walsh, G. and Murphy, B., Eds., Dordrecht: Kluwer Academic Publisher, pp. 109–123, 1999.

Harousseau, J.L. et al., Granulocyte colony-stimulating factor after intensive consolidation chemotherapy in acute myeloid leukemia: results of a randomized trial of the Groupe Ouest-Est Leucemies Aigues Myeloblastiques, *J. Clin. Oncol.*, 18, 780–787, 2000.

Herman, A.C. et al., Characterization, formulation, and stability of Neupogen (Filgrastim), a recombinant human granulocyte-colony stimulating factor, *Pharm. Biotechnol.*, 9, 303–328, 1996.

Norditropin

Product Name: Norditropin (trade name)

Somatropin (rDNA origin) (common name)

Description: Norditropin is a highly purified preparation of human growth hormone (hGH) produced by recombinant DNA technology. The 191-amino acid, single-chain, 22,124-Da polypeptide is identical to native hGH. It is available both in freeze-dried and liquid formulations. The lyophilized format is available in various final product configurations, either in vials containing 4 or 8 mg active ingredient or in cartridges for use in conjunction with a pen-like autoinjector administration device (the NordiPen). The diluent provided is usually water for injection (WFI) containing 1.5% benzyl alcohol as a preservative, although the product can be reconstituted in WFI alone for patients who are allergic to benzyl alcohol. The liquid formulation of the product is referred to as Norditropin SimpleXx. The product is stored at 2 to 8°C and is normally administered at concentrations of 2-4 mg active/ml subcutaneously.

Approval Date: 1995 (U.S.; although marketing was delayed by 2 years due to a patent dispute)

Therapeutic Indications: Norditropin is indicated for the long-term treatment of children who have growth failure due to inadequate secretion of endogenous growth hormone.

Manufacturer: Novo Nordisk A/S 2880 Bagsvaerd, Denmark

Marketing: Novo Nordisk Pharmaceuticals Inc., 100 Overlook Center, Suite 200, Princeton, NJ, 08540, http://www.novonordisk-us.com (U.S.)

Manufacturing

Norditropin is produced by recombinant DNA technology in an engineered strain of *Escherichia coli*. The producer carries a copy of the native hGH cDNA, placed under an *E. coli* signal sequence, in a circular plasmid. Batch production is initiated by fermentation of the producer *E. coli* cell line. Purification entails a combination of high-resolution chromatographic steps. This is followed by product formulation, sterile filtration, and lyophilization, when the product is sold in freeze-dried format. Added excipients include glycine, sodium phosphate buffer, and mannitol.

Overview of Therapeutic Properties

Norditropin is a potent metabolic hormone that displays biological activities identical to native endogenous growth hormone. It influences the metabolism of lipids, carbohydrates, and protein. It maintains normal body composition in adults and children by increasing nitrogen retention and the stimulation of skeletal muscle growth, as well as by mobilizing body fat. hGH stimulates skeletal growth mainly by increasing levels of insulin-like growth factor-1, and clinical trials have directly demonstrated that administration of Norditropin is associated with improvements in the predicted heights achieved in growth hormone-deficient children. The incidence of drug-related adverse effects are low, and the product is contraindicated in patients with evidence of active malignant tumors.

Further Reading

Drent, M.L. et al., Acceptability of liquid human growth hormone (hGH) [Norditropin Simplex Xx ®] in adults and children with GH deficiency and children with chronic renal disease, *Clin. Drug Invest.*, 22(9), 633–638, 2002.

Garcia, J.J. et al., Biodegradable laminar implants for sustained release of recombinant human growth hormone, *Biomaterials*, 23(24), 4759–4764, 2002.

Laursen, T. et al., Pharmacokinetic and metabolic effects of growth hormone injected subcutaneously in growth hormone deficient patients. Thigh versus abdomen, *Clin. Endocrinol.*, 40(3), 373–378, 1994.

Stanhope, R. et al., An open label acceptability study of Norditropin Simplex Xx — a new liquid growth hormone formulation, *J. Pediatr. Endocr. Met.*, 14(6), 735–740, 2001.

Novolin/Actrapid/Insulatard/Mixtard/ Monotard/Ultratard/Velosulin

Product Name: Novolin/Actrapid/Insulatard/Mixtard/Monotard/Ultratard/Velosulin (trade name)

Human insulin, rDNA (international nonproprietary name)

Description: The recombinant human insulin from Novo Nordisk is produced in *Saccharomyces cerevisiae* using recombinant DNA technology. It is supplied in numerous formulations that exhibit different onsets of action and duration. Actrapid (Novolin R in the U.S.) is a fast-acting insulin that consists of zinc insulin prepared for subcutaneous and intravenous administration. Insulatard (also marketed with the trade name Protaphane and with the trade name Novolin N [NPH] in the U.S.) is a long-acting insulin that consists of a suspension of isophane insulin for subcutaneous administration. Mixtard (also marketed with the trade name Actraphane and with the trade name Novolin 70/30 in the U.S.) is a dual-acting insulin consisting of a suspension of a different combination of dissolved insulin and isophane insulin for subcutaneous administration. Monotard (Novolin L or Lente in the U.S.) is an intermediate-acting insulin that consists of a suspension of zinc insulin, 30% amorphous and 70% crystalline, for subcutaneous administration. Ultratard (Novolin Ultralente in the U.S.) is a long-acting insulin consisting of a suspension of crystalline particles of zinc insulin for subcutaneous administration. Velosulin is a fast-acting insulin consisting of a solution for continuous subcutaneous infusion with infusion pumps for intravenous and intramuscular injection. All the products are supplied in vials, while Actrapid, Insulatard, and Mixtard are also available in cartridges that are used with dispenser devices (Penfill) and in prefilled syringes (NovoLet, InnoLet, and FlexiPen).

Approval Date:	2002 (E.U.); 2001–2003 (U.S.). All products are also approved in several other countries. Some Novolin formulations were withdrawn from the U.S. market in 2003.
Therapeutic Indications:	All formulations of the recombinant human insulin are used for the treatment of patients suffering from diabetes mellitus.
Manufacturer:	Novo Nordisk A/S/, Novo Allé, 2880 Bagsvaerd, Denmark and (for Penfill only) Novo Nordisk Pharmaceutique S.A., 45 Avenue d'Orléans, 28002 Chartres, France, http://www.novonordisk.com (manufacturers are responsible for batch release in the European Economic Area) Novo Nordisk Pharmaceuticals, Inc., 100 College Road West, Princeton, NJ 08540, http://www.novonordisk-us.com (U.S.)
Marketing:	Novo Nordisk A/S/, Novo Allé, 2880 Bagsvaerd, Denmark, http://www.novonordisk.com (E.U.) Novo Nordisk Pharmaceuticals, Inc., 100 College Road West, Princeton, NJ 08540, http://www.novonordisk-us.com (U.S.)

Manufacturing

The recombinant insulin is produced in a genetically engineered strain of *S. cerevisiae* harboring a recombinant plasmid construct based on the yeast 2μ plasmid. The plasmid houses a sequence coding for a single amino acid chain insulin precursor, consisting of the mature insulin's A and B chains linked by a 3-amino acid linker region. This is attached to a pre-pro leader region of the yeast-mating factor gene, MFα1. During fermentation, the insulin precursor molecule is secreted into the extracellular fermentation medium, from where it is recovered and further processed. The crude insulin precursor is initially converted into the mature insulin methyl ester by treatment with trypsin (derived from porcine pancreas). This is then subjected to hydrolysis, yielding the mature insulin. Purification is achieved by a number of chromatographic and precipitation steps. Approprite excipients are then added and the product is filter sterilized and filled into final containers.

The purified product is presented in different formulations. Actrapid is presented as a solution for subcutaneous and intravenous administration and contains dissolved zinc insulin crystals. Insulatard consists of insulin crystallized with zinc and presented in suspension with protamine sulphate (isophane insulin) for subcutaneous administration. Mixtard is a combination of two forms of insulin — dissolved zinc insulin and insulin isophane — for subcutaneous administration; it is available in different combinations of the two forms, containing 10, 20, 30, 40, or 50% of dissolved insulin, with the remainder made of isophane insulin. In the U.S., it is available as Novolin 70/30 in the form of a solution containing 70% isophane insulin and 30% dissolved insulin. Monotard is a suspension containing a mixture of insulin in an amorphous form (30%) and as crystalline particles (70%) for subcutaneous administration. Ultratard is formulated as a suspension of crystalline particles of insulin and is used for subcutaneous administration. Velosulin is presented as a solution of dissolved insulin for continuous infusion using infusion pumps.

Actrapid and Velosulin contain insulin, zinc chloride, glycerol, metacresol, sodium hydroxide, and hydrochloric acid. Insulatard and Mixtard contain insulin, zinc chloride, glycerol, metacresol, phenol, dihydrate disodium phosphate, sodium hydroxide, hydrochloric acid, and protamine sulphate. Ultratard contains insulin, zinc chloride, sodium chloride, sodium acetate, methyl parahydroxybenzoate, sodium hydroxide, and hydrochloric acid. Monotard contains insulin, zinc chloride, zinc acetate, sodium chloride, sodium acetate, methyl parahydroxybenzoate, sodium hydroxide, and hydrochloric acid.

All products are supplied in vials, while Actrapid, Insulatard, and Mixtard are also available in cartridges, which are used with dispenser devices (Penfill) and in prefilled syringes (NovoLet, InnoLet, and FlexiPen). The products have a shelf life of 30 months when stored at 2 to 8°C in light-protected containers.

Overview of Therapeutic Properties

The recombinant insulin is very similar to the natural molecule and is indicated for the treatment of patients with diabetes mellitus. The various formulations differ in their onset of action and duration of activity. Actrapid and Velosulin are fast-acting insulin with an activity similar to the natural molecule. They each exhibit an onset of action within 30 minutes of administration, a maximum effect between 90 and 210 minutes, and a duration of 7 to 8 hours. Actrapid is generally administered subcutaneouly but may also be administered intravenously. It is often administered in combination with a long-acting insulin. Velosulin is used for continuous subcutaneous infusion with external infusion pumps and for intramuscular or intravenous injection, with half of the daily dose given as a continuous basal rate and the other half given in bolus form during meal times. Insulatard is a long-acting insulin with an onset of action within 90 minutes of administration, a maximum effect between 4 and 12 hours from administration, and a duration of up to 24 hours. Insulatard may be administered in combination with a fast-acting insulin. Mixtard is a dual-acting insulin containing a fast-acting and a long-acting form of insulin. It is available in various formulations containing different amounts of the fast-acting, dissolved form of insulin and the long-acting isophane insulin form. For the U.S. market, Novolin 70/30 contains 70% of isophane insulin and 30% of dissolved insulin. The onset of action of Mixtard occurs within 30 minutes of administration, with a maximum effect observed after 2 to 8 hours, and a duration of 24 hours. Monotard is an intermediate-acting insulin consisting of a combination of two different forms of insulin — the amorphous form and crystalline particles — that exhibit an onset of action 150 minutes after administration, a maximum effect after 7 to 15 hours, and a duration of action of 24 hours. Ultratard is another long-acting insulin with an even longer duration than Monotard. It consists of a suspension of crystalline particles of zinc insulin with an onset of action observed within 4 hours of administration, a maximum effect between 8 and 24 hours, and a duration of 28 hours. Insulatard, Mixtard, Monotard, and Ultratard are administered subcutaneously once or twice a day. Required dosages are determined based on the individual needs of the patient.

All the various formulations of recombinant human insulin were found to be safe and well tolerated. Hypoglycemia was the most commonly observed side effect. Reactions and lipodystrophy at the site of injection, allergic reactions, visual disorders, and edema were also reported, though rarely. These products are contraindicated in the case of hypoglycemia.

Further Reading

http://www.eudra.org
http://www.fda.gov
http://www.novonordisk.com
Aronoff, S. et al., Use of a premixed insulin regimen (Novolin 70/30) to replace self-administered insulin regimes, *Clin. Ther.*, 16(1), 41–49, 1994.
Plevin, S. and Sadur, C., Use of a prefilled insulin syringe (Novolin prefilled TM) by patients with diabetes, *Clin. Ther.*, 15(2), 423–431, 1993.

NovoRapid/Novolog and NovoMix 30/ NovoLog Mix 70/30

Product Name: NovoRapid/Novolog and NovoMix 30/NovoLog Mix 70/30 (trade names)

Insulin aspart (international nonproprietary name)

Description: NovoRapid (Novolog in the U.S.) is a fast-acting human insulin analogue. It differs from the human insulin in an aspartic acid residue, replacing a proline at position 28 in the B chain. NovoRapid is produced by recombinant DNA technology in a *Saccharomyces cerevisiae* strain, and it is presented as a solution (100 IU/ml) for subcutaneous administration. NovoMix 30 (NovoLog Mix 70/30 in the U.S.) is a new formulation containing 30% soluble insulin aspart and 70% protamine-crystallized insulin aspart, resulting in an intermediate-acting insulin analogue.

Approval Date: 1999 (E.U., for NovoRapid); 2000 (U.S., for NovoLog); 2001 (U.S. only, for the use of NovoLog in insulin external pumps); 2000 (E.U., for NovoMix 30); 2002 (U.S. for NovoLog Mix 70/30).

Therapeutic Indications: Diabetes mellitus.

Manufacturer: Novo Nordisk A/S, Novo Allé, 2880 Bagsvaerd, Denmark, http://www.novonordisk.com (manufacturer is responsible for batch release in the European Economic Area)
Novo Nordisk Pharmaceuticals, Inc., 100 College Road West, Princeton, NJ 08540, http://www.novonordisk-us.com (manufacturer responsible for import and batch release in the U.S.)

Marketing: Novo Nordisk A/S, 2880 Bagsvaerd, Denmark, http://www.novonordisk.com (E.U.)
Novo Nordisk Pharmaceuticals, Inc., 100 College Road West, Princeton, NJ 08540, http://www.novonordisk-us.com (U.S.)

Manufacturing

Insulin aspart is a recombinant human insulin analogue. It is identical to human insulin except for an aspartic acid residue replacing a proline at position 28 in the B chain. The protein is produced by recombinant DNA technology in a modified strain of *S. cerevisiae*. Following recovery from the culture medium, the protein is subjected to a number of high-resolution chromatographic steps and viral removal procedures. NovoRapid consists of the active substance insulin aspart and the following excipients: glycerol, phenol, metacresol, zinc chloride, sodium chloride, disodium dihydrate phosphate, sodium hydroxide, and hydrochloric acid. NovoMix 30 contains 30% of insulin aspart in soluble form and 70% crystallized with protamine sulphate, as well as the following excipients: mannitol, phenol, metacresol, zinc chloride, sodium chloride, disodium dihydrate phosphate, sodium hydroxide, and hydrochloric acid. The products are provided as a solution (or a suspension in the case of NovoMix 30) for subcutaneous administration in vials, prefilled sterile syringes, and cartridges for use in a prefilled multidose device.

The shelf life of both products is 24 months when stored at 2 to 8°C, protected from direct light. Extensive tests are carried out to ensure the quality and safety of the final products.

Overview of Therapeutic Properties

NovoRapid is a fast-acting analogue of human insulin. It is indicated for the treatment of patients with diabetis mellitus. The amino acid modification of insulin aspart decreases the tendency of the molecule to form hexamers; therefore, it has the advantage over human insulin of being adsorbed more rapidly, resulting in a more rapid onset of action. NovoRapid can be taken just before meals, unlike human insulin which needs to be taken 30 minutes beforehand. This provides more flexibility for patients. Its maximum effect is attained between 1 and 3 hours after administration, and it lasts for up to 5 hours. NovoRapid has a shorter duration of action than human insulin, and intermediate-action or long-action insulin formulations should be administered at least once a day. A new formulation of the product is now available (NovoMix 30), in which 30% of the insulin aspart exists in a soluble form and 70% in a crystallized form. The crystallized form has a slower absorption, similar to the human insulin; thus, NovoMix 30 is a longer-acting product, with an effect that lasts for up to 24 hours after injection. NovoRapid and NovoMix 30 are administered subcutaneously with rotation of the site of injection.

NovoRapid and NovoMix 30 proved to be efficient in controlling blood glucose levels in type 1 and type 2 diabetic patients. No major side effects were observed after administration of the products; however, hypoglycemia, which is commonly described with the use of insulin and insulin analogues, was occasionally reported. Local allergic reactions have also been reported. Antibodies against insulin aspart have developed in the first 6 months after administration, but after 1 year they had reduced to an insignificant titre. Insulin aspart is contraindicated in the case of hypoglycemia.

Further Reading

http://www.eudra.org
http://www.fda.gov
http://www.novolog.com
http://www.novonordisk.com
http://www.novorapid.com
Chapman, T.M. et al., Insulin aspart: a review of its use in the management of type 1 and 2 diabetes mellitus, *Drugs*, 62, 1945–1981, 2002.
Setter, S.M. et al., Insulin aspart: a new rapid-acting insulin analog, *Ann. Pharmacother.*, 34, 1423–1431, 2000.

Novoseven

Product Name:	Novoseven (trade name) Eptacog alfa (international nonproprietary name)
Description:	Novoseven is a recombinant human coagulation factor VIIa (rFVIIa), a serine protease involved in the coagulation cascade that terminates with the production of a fibrin clot. It is produced in baby hamster kidney (BHK) cells and is supplied in a lyophilized form to be reconstituted in water for injections, usually to a concentration of 0.6 mg/ml active ingredient (equivalent to 30 IU/ml activity). The product is administered by intravenous injection.
Approval Date:	1996 (E.U.); 1999 (U.S.)
Therapeutic Indications:	Novoseven is indicated for the treatment of bleeding episodes and surgery in patients with hemophilia A or B, with inhibitors to coagulation factor VIII or factor IX.
Manufacturer:	Novo Nordisk A/S Hagedornsvej, 2820 Gentofte, Denmark, http://www.novonordisk.com (manufacturer is responsible for batch release in the European Economic Area) Novo Nordisk A/S, Novo Alle, 2880 Bagsvaerd, Denmark, http://www. novonordisk.com (U.S.)
Marketing:	Novo Nordisk A/S, 2880 Bagsvaerd, Denmark, http://www.novonordisk.com (E.U.) Novo Nordisk Pharmaceuticals, Inc., 100 College Road West, Princeton, NJ 08540, http://www.novonordisk-us.com (U.S.)

Manufacturing

rFVII gene was cloned and is expressed in BHK cells. The recombinant FVII is produced as a single chain 50 KDa, 406 amino acid glycoprotein. This is converted by autocatalysis into the active two-chain form (FVIIa) during chromatographic purification — a 20-kDa light chain and a 30-kDa heavy chain connected by a single disulphide bond. The enzyme's catalytic site resides on the heavy chain.

The purification process includes affinity and ion exchange chromatography and procedures to ensure viral safety. The final product is presented in a lyophilized form consisting of eptacog alfa (activated), as well as sodium chloride, calcium chloride dihydrate, glycylglycine, polysorbate 80, and mannitol as excipients. It contains traces of mouse IgG (from the affinity chromatography purification) and bovine and hamster proteins (from culture serum).

Extensive control tests carried out during the processing and on the final product to ensure quality and safety include mass spectroscopy, amino acid analysis, tryptic mapping, carbohydrate analysis, circular dichroism spectra, SDS-PAGE, Western blotting, and RP-HPLC.

Overview of Therapeutic Properties

Novoseven is used to stimulate the coagulation process in hemophilic patients with inhibitors to factors VIII and IX. Novoseven induces the activation of factor X into factor Xa, which converts prothrombin into thrombin, leading to the final stage where fibrinogen is converted into fibrin to form the hemostatic plug; the whole process achieves by-pass of the action of factors VIII and IX. This activation occurs only in the presence of tissue factor (a membrane protein not present in the plasma), calcium, and phospholipids, so that coagulation is stimulated only when an injury has occurred to a vessel, with resulting local hemostasis.

Novoseven proved to be effective and safe in the treatment of bleeding episodes. It is administered to patients every 2 to 3 hours until hemostasis is obtained. In the case of mild bleeds treated at home (joint, muscle, and mucocutaneous bleeds), three doses of Novoseven are given to stop the bleeding, and one additional dose is given to maintain hemostasis. For serious bleeds treated in the hospital, the treatment could continue for up to a maximum of 2 to 3 weeks. In the case of surgery, an initial dose is given immediately before operation.

Generally mild side effects were observed in patients after Novoseven treatment, and no human antibodies against FVIIa have been detected. The potential risk of thrombotic events or disseminated intravascular coagulation has been reported in patients with a high amount of tissue factor circulating in blood. Caution needs to be taken for administration of Novoseven during pregnancy and lactation.

Further Reading

http://www.eudra.org
http://www.fda.gov
http://www.novonordisk.com
http://www.novoseven.com
Hay, C.R. et al., The treatment of bleeding in acquired haemophilia with recombinant factor VIIa: a multicentre study, *Thromb. Haemost.*, 78, 1463–1467, 1997.
Scharrer, I., Recombinant factor VIIa for patients with inhibitors to factor VIII or IX or factor VII deficiency, *Haemophilia*, 5, 253–259, 1999.

Nutropin Depot

Product Name:	Nutropin Depot (trade name) Somatropin (international nonproprietary name)
Description:	Nutropin Depot is a different formulation of the product Nutropin (lyophilized form) or Nutropin-Aq (soluble form). Nutropin Depot is a slow-release, long-acting microcapsulated form of the recombinant human growth hormone (somatropin). It is identical in amino acid sequence and activity to the natural 191 amino acid, 22-kDa human hormone. Nutropin Depot is produced by recombinant DNA technology in *Escherichia coli* and is presented as a suspension for subcutaneous injection.
Approval Date:	1999 (U.S.)
Therapeutic Indications:	Nutropin Depot is indicated for the treatment of children with growth failure due to insufficient endogenous growth hormone.
Manufacturer:	Genentech, Inc., 1 DNA Way, South San Francisco, CA, http://www.gene.com (U.S.)
Marketing:	Genentech, Inc., 1 DNA Way, South San Francisco, CA, http://www.gene.com (U.S.)

Manufacturing

Nutropin Depot contains the same somatropin human recombinant growth hormone as Nutropin and NutropinAq, presented as a different formulation. The recombinant hormone is produced in *E. coli* using recombinant DNA technology. A bacterial signal sequence is added in front of the hormone nucleotide sequence in order to avoid addition of a methionine residue at the protein N-terminus during production in the bacterium. The expressed protein is transported into the *E. coli* periplasmic space, during which the additional N-terminal sequence is removed, resulting in a functional recombinant protein.

The protein undergoes multiple chromatographic purification steps and procedures for the removal of viral and bacterial contaminants. The purified product is lyophilized in the presence of stabilizing agents and embedded at low temperature in biodegradable polylactide-coglycolide polymer microspheres. Nutropin Depot consists of microencapsulated lyophilized powder of the human recombinant growth hormone, zinc acetate, zinc carbonate, and polylactide-coglycolide and is supplied with a solution containing carboxymethylcellulose sodium salt, polysorbate 20, and sodium chloride to be used for dilution before subcutaneous injection.

Overview of Therapeutic Properties

Nutropin Depot is indicated in the treatment of children who suffer from growth failure due to a deficiency of endogenous growth hormone. Nutropin Depot exhibits the same activity as Nutropin and NutropinAq in terms of stimulation of skeletal growth. Unlike Nutropin and NutropinAq, Nutropin Depot is encapsulated in and slowly released from biodegradable microspheres, resulting in a long-term action with administration and thus required only once or twice in a month as opposed to daily (in the case of Nutropin and NutropinAq). Nutropin Depot proved to be effective in stimulating the skeletal growth, with up to a few centimeters of growth per year reported.

Nutropin Depot was found to be safe, with reactions at the injection site and headaches being the most commonly reported side effects. The development of antisomatropin antibodies was also reported. Nutropin Depot is contraindicated in patients with closed epiphyses, active neoplasia, acute respiratory failure, and illness due to complications after open heart surgery, abdomen surgery, or multiple accidental trauma.

Further reading

http://www.fda.gov
http://www.gene.com
Cook, D.M. et al., The pharmacokinetic and pharmacodynamic characteristics of a long-acting growth hormone (GH) preparation (Nutropin Depot) in GH-deficient adults, *J. Clin. Endocrinol. Metab.*, 87, 4508–4514, 2002.
Johnson, B.C. and Barcia, J.P., Use of Nutropin Depot in short children with chronic renal insufficiency, *Pediatr. Res.*, 53(4), 3010, 2003.

Rincon, M. et al., Comparison of the efficacy and safety of weekly doses of long acting growth hormone (GH) formulations (Nutropin Depot) to daily GH Nutropin AQ, *Pediatr. Res.*, 53(4), 847, 2003.

Silverman, B.L. et al., A long-acting human growth hormone (Nutropin Depot): efficacy and safety following two years of treatment in children with growth hormone deficiency, *J. Pediatr. Endocrinol. Metab.*, 15(Suppl. 2), 715–722, 2002.

NutropinAQ

Product Name: NutropinAq (trade name)

Somatropin (international nonproprietary name)

Description: NutropinAq is an aqueous formulation of (recombinant) human growth hormone (hGH) (somatropin), a 191-amino acid, 22-kDa polypeptide hormone. NutropinAq is produced by recombinant DNA technology in a modified *Escherichia coli* strain. It is presented as a solution for subcutaneous administration in vials (or, in some world regions, in cartridges to be used with a delivery device, the NutropinAQ Pen).

Approval Date: 2001 (E.U.); 1995 (U.S.). NutropinAq is also approved in several other countries.

Therapeutic Indications: NutropinAq is indicated for the treatment of children with growth failure due to insufficient endogenous growth hormone or due to Turner's syndrome. It is also indicated for prepubertal children with growth failure due to chronic renal insufficiency or prior renal transplantation and for adults with a growth hormone deficiency.

Manufacturer: Schwarz Pharma A.G., Alfred Nobel Strasse 10, 40789 Monheim, Germany, http://www.schwarzpharma.com (manufacturer is responsible for batch release in the European Economic Area)
Genentech, Inc., 1 DNA Way, South San Francisco, CA 94080-4990, http://www.gene.com (U.S.)

Marketing: Schwarz Pharma A.G., Alfred Nobel Strasse 10, 40789 Monheim, Germany, http://www.schwarzpharma.com (E.U.)
Genentech, Inc., 1 DNA Way, South San Francisco, CA 94080-4990, http://www.gene.com (U.S.)

Manufacturing

NutropinAq is a recombinant hGH. It consists of a peptidic chain with two intrachain disulphide bridges, very similar to the natural growth hormone. It is produced by recombinant DNA technology in *E. coli* cells. The peptide is expressed in bacterial cells and transported into the periplasmic space, due to the addition of an N-terminal sequence that is then removed. This process results in the production of a molecule identical to the natural human hormone, without the addition of the N-terminal methionine residue, which is present in several other recombinant hGH products. The molecule is extensively purified using several chromatographic steps, including gel filtration, ion exchange and immunoaffinity chromatography, and filtration techniques. The final product is presented as a solution for subcutaneous administration in vials (or in cartridges for use with a delivery device, the NutropinAQ Pen, in some world regions). NutropinAq consists of the active substance somatropin, sodium chloride (as a solubility and tonicity agent), phenol (as a preservative), polysorbate 20 (as a solubilizer), sodium citrate, and citric acid (as a buffering agent).

The shelf life of the product is 18 months when stored at 2 to 8°C. The quality and the safety of the products are ensured by testing to determine its purity, efficacy, potency, and viral and bacterial safety.

Overview of Therapeutic Properties

NutropinAq is indicated in the treatment of children who suffer from growth failure due to lack of endogenous growth hormone, Turner's syndrome, renal insufficiency, prior renal transplantation, during prepubertal period, and also in the treatment of adults who suffer from growth hormone deficiency. NutropinAq exhibits the same activity as the natural growth hormone in stimulation of bone growth and results in increased height in children of a few centimeters after 4 to 5 years of treatment. In adults, reduction in the fat mass and increase in spinal bone mineral density and lean mass were observed.

NutropinAq should be administered subcutaneously once a day at the dosage required for the age group, and rotating the site of injection. NutropinAq proved an effective replacement of the human growth hormone. It showed efficacy equivalent to Nutropin, a product previously marketed in the U.S. that is identical to NutropinAq but is presented in a lyophilized form.

NutropinAq was shown to be safe, with reactions at the injection site and headaches being the most commonly reported side effects. The development of antisomatropin antibodies was observed, but with no effect on the efficacy of the treatment. NutropinAq is contraindicated during pregnancy and lactation and in patients with closed epiphyses, active neoplasia, acute respiratory failure, or illness due to complications after open heart surgery, abdomen surgery, or multiple accidental trauma.

Further Reading

http://www.eudra.org
http://www.fda.gov
http://www.gene.com
See also Further Reading section of the *Nutropin Depot* monograph.

OncoScint CR/OV

Product Name: OncoScint CR/OV (trade name)

Satumomab pendetide (common name)

Description: The active ingredient in OncoScint is a murine monoclonal antibody conjugated to a linker chelator molecule (glycyl-tryosyl-N-diethylenetriaminepentacetic acid; GYK-DTPA). The antibody is of the IgG1κ subclass and has been raised against a high molecular weight tumor associated glycoprotein (TAG-72), which is often associated with ovarian or colorectal cancer. The chelator is covalently attached to the antibody such that it can bind radioactive Indium (^{111}In) prior to its administration for *in vivo* radiological-based diagnostic imaging (immunoscintigraphy). The product is supplied as a 2-vial kit containing all the nonradioactive ingredients required to generate a single-unit dose of OncoScint. Vial 1 contains 1 mg conjugated antibody in a final volume of 2 ml phosphate buffered saline, pH 6.0. Vial 2 contains 2 ml of sodium acetate buffer, pH 6.0. The kit also contains a 0.22-μm filter.

Approval Date: 1992 (U.S.)

Therapeutic Indications: This product is used as a diagnostic imaging agent (after conjugation to radioactive Indium) for determining the extent and location of cancer tissue in patients with proven colorectal or ovarian cancer.

Manufacturer: Cytogen Corporation, 650 College Road East, Suite 3100, Princeton, NJ, http://www.cytogen.com

Marketing: Cytogen Corporation, 650 College Road East, Suite 3100, Princeton, NJ, http://www.cytogen.com

Manufacturing

The murine monoclonal antibody is derived from a hybridoma cell line. The antibody is produced by culture of the hybridoma cells in an air lift bioreactor. Subsequent to removal of cell mass the antibody is purified from the extracellular media using a combination of high-resolution chromatographic procedures. A viral inactivation step is also included in the downstream processing procedure. The chelator (GYK-DTPA) is then covalently attached to the antibody (via the latter's oligosaccharide side chains). The antibody's carbohydrate moieties are first chemically oxidized, rendering them reactive with the GYK-DTPA. Unreacted elements are removed from the antibody conjugate by gel filtration chromatography. The product is then filter sterilized and aseptically filled into sterile 6-ml glass vials.

Overview of Therapeutic Properties

In vitro investigations indicate that OncoScint will bind to more than 80% of colorectal adenocarcinomas and in excess of 90% of common epithelial ovarian carcinomas. It will also bind to a high proportion of breast, gastric, esophageal, pancreatic, and small cell lung cancers. In contrast, the antibody reacts poorly with and is unreactive against most healthy human cells and tissue. Immediately prior to product administration, ^{111}In (not provided with the kit) is chelated to the antibody conjugate. ^{111}In emits γ-radiation (with a half-life of 2.8 days), which can be detected using a gamma camera. The usual product dose is 1 mg OncoScint radiolabeled with 5 mCi of ^{111}In. Good localization to primary and metastatic tumor sites is generally observed. However, localization of radioactivity in the liver, spleen, and bone marrow is also observed. This is likely to be largely and exclusively due to removal and metabolism of the conjugate at these sites. Administration of OncoScint induces a HAMA response (production of human antimouse antibodies) in 55% of patients after a single product dose.

Further Reading

Divgi, C.R., Radiolabelled monoclonal antibody imaging in cancer — potential and limitations, *Clin. Immunother.*, 3(3), 218–226, 1995.

Nabi, H.A., The use of radiolabelled monoclonal antibodies in the diagnosis of colorectal carcinomas, *Clin. Immunother.*, 6(3), 200–210, 1996.

Neal, C.E. et al., Quantitative analysis of In-111 satumomab pendetide immunoscintigraphy — an aid to visual interpretation of images in patients with suspected carcinomatosis, *Clin. Nucl. Med.*, 21(8), 638–642, 1998.
Peters, D. and Fitton, A., Satumomab pendetide — a prelimiary review of its use in the diagnosis of colorectal and ovarian cancer, *Clin. Innumother.*, 3(5), 395–408, 1995.

Ontak

Product Name: Ontak (trade name)

Denileukin diftitox (international nonproprietary name)

Description: Denileukin diftitox is an engineered recombinant protein produced in *Escherichia coli* using recombinant DNA technology. It consists of a modified form of diphtheria toxin fused to human interleukin 2 (IL-2). Specifically, this 58-kDa protein consists of a portion of diphtheria toxin fragments A and B from *Corynebacterium diphtheriae* strain C7 fused to human IL-2. The amino acid sequence consists of an N-terminal methionine residue, amino acids 1–386 and 484–485 from the diphtheria toxin, and amino acids 2–133 of IL-2. The eukaryotic binding site of the diphtheria toxin has been deleted in the fusion protein, but the cytotoxic and the translocation domains of the toxin are retained. It is provided as a frozen solution (150 μg/ml active ingredient) to be administered intravenously.

Approval Date: 1999 (U.S.)

Therapeutic Indications: Ontak is indicated for the treatment of adult patients with persistent or recurrent cutaneous T-cell lymphoma whose malignant cells express the CD25 component of the IL-2 receptor.

Manufacturer: Seragen, Inc., 97 South Street, Hopkinton, MA 01748, http://www.seragen.com (U.S.)

Marketing: Ligand Pharmaceuticals, 10275 Science Center Drive, San Diego, CA 92121, http://www.ligand.com (U.S.)

Manufacturing

Denileukin diftitox is a fusion protein produced in *E. coli* using recombinant DNA techniques. The fusion protein is expressed in the *E. coli* culture under the control of an inducible promoter. Once the culture reaches a specific cell density, expression of the fusion protein is induced for 2 hours, followed by harvesting and lysis of cells. Purification involves several chromatographic steps, including immunoaffinity chromatography using diphtheria toxin antibody, HPLC, reverse phase chromatography, and filtration procedures. The final product is available as a frozen solution for intravenous administration and contains denileukin diftitox, as well as citric acid, ethylenediamine tetraacetic acid, and polysorbate as excipients.

The shelf life of the product is 12 months when stored at a temperature below –10°C. Extensive tests are carried out to ensure the product's quality and safety.

Overview of Therapeutic Properties

Cutaneous T-cell lymphoma (CTCL) is a form of non-Hodgkin's lymphoma. It appears initially in the skin and slowly spreads to the lymph nodes, spleen, liver, and other organs, with a median survival ranging from less than 3 years (when organs are involved) to greater than 10 years, depending on the stage of the disease at the time of diagnosis. As in the case of leukemias and other lymphomas, malignant cells of the CTCL express IL-2 receptor, which is usually found on activated T cells, B cells, and macrophages. The fusion protein denileukin diftitox, the active substance of Ontak, allows targeting of CTCL cells by directing the diphtheria toxin to cells expressing the IL-2 receptor (CD25). After binding to the receptor, the protein is internalized, and the toxin induces cell death by inhibiting protein synthesis. Denileukin diftitox also binds some normal lymphocytes.

Ontak is used in patients with persistent or recurrent CTCL whose malignant cells express the CD25 receptor. It should be administered intravenously daily for 5 days every 3 weeks. Studies were performed on patients who had not responded to an average of five previous treatments and who received a median of six courses of therapy. A partial response was obtained in 20% of patients treated with Ontak, while a complete response was obtained in 10 to 15% of patients in whom remission had not previously been achieved with other drugs. Overall, patients gained 10 or more years of survival after Ontak treatment.

The most commonly reported side effects were infections, allergic reactions, and vascular leak syndrome, which were severe and life threatening in 5% of cases. Antibodies against diphtheria toxin developed in almost all patients after a few courses of therapy, probably due to prior vaccination against diphtheria toxoid. Antibodies against IL-2 were reported in 50% of patients. Ontak should be administered only when necessary during pregnancy, and it is contraindicated during lactation.

Further Reading

http://www.fda.gov
http://www.ligand.com
Duvic, M. et al., Quality-of-life improvements in cutaneous T-cell lymphoma patients treated with denileukin diftitox (ONTAK(R)), *Clin. Lymphoma*, 2, 222–228, 2002.
Foss, F.M., Interleukin-2 fusion toxin: targeted therapy for cutaneous T-cell lymphoma, *Ann. N. Y. Acad. Sci.*, 941, 166–176, 2001.
Olsen, E. et al., Pivotal phase III trial of two dose levels of denileukin diftitox for the treatment of cutaneous T-cell lymphoma, *J. Clin. Oncol.*, 19, 376–388, 2001.

Optisulin

Product Name: Optisulin (trade name)

Insulin glargine (international nonproprietary name)

Description: Optisulin is a long-acting human insulin analogue. It is apparently identical to the product marketed with the trade name Lantus. It differs from the human insulin at the C-terminal end of the B chain, which is elongated by two arginine residues, and at the C terminus of the A chain, where an asparagine is replaced by a glycine. Optisulin is produced in modified *Escherichia coli* by recombinant DNA technology and is presented as a solution for subcutaneous injection.

Approval Date: 2000 (E.U.)

Therapeutic Indications: Optisulin is indicated for patients with diabetes mellitus where insulin treatment is required.

Manufacturer: Aventis Pharma Deutschland GmbH, Brünistrasse 50, 65926 Frankfurt am Main, Germany, http://www.aventis.com (manufacturer is responsible for batch release in the European Economic Area)

Marketing: Aventis Pharma Deutschland GmbH, Brünistrasse 50, 65926 Frankfurt am Main, Germany, http://www.aventis.com

Manufacturing

Insulin glargine is a recombinant human insulin analogue. It is produced by recombinant DNA technology in *E. coli* following cloning of the modified human insulin chains. The purification process involves a number of chromatographic steps followed by precipitation. The final product contains the active ingredient insulin glargine and the following excipients: zinc-chloride (as a stabilizer), m-cresol (as a preservative), glycerol (as a tonicity agent), hydrochloric acid (to dissolve the active ingredient), and sodium hydroxide (to adjust the pH). It is presented as a solution with a pH of 4.0 in vials or in cartridges to be used with multidose devices for subcutaneous injection.

The shelf life of the product is 24 months when stored at 2 to 8°C. Extensive tests have been carried out to ensure the quality and the safety of the product.

Overview of Therapeutic Properties

Optisulin is a long-acting human insulin analogue. The genetically modified molecule has an isoelectric point that differs from that of natural human insulin, resulting in a molecule that is more soluble in an acid environment and less stable at physiological pH. When the product is injected into a neutral environment, it forms microprecipitates, from which insulin is released slowly over a long period of time. Administration of optisulin is not followed by the usual activity peak that is observed with native human insulin, and its action continues at a constant rate for up to 24 hours. Optisulin is indicated for patients with diabetes mellitus, where insulin is required. The dosage should be adjusted for the individual needs of the patients, and it should be administered once daily in the evening with rotation of the injection site.

Optisulin proved its efficacy in controlling blood glucose levels in patients with type 1 and type 2 diabetes mellitus. The most common observed side effects are those commonly reported for insulin and other insulin analogues, including hypoglycemia, immunoreactions, development of antibodies, and visual disorders. Pain and reactions at the sites of injection are observed more pronouncedly when optisulin is used compared to other insulin products.

No studies have been carried out on children. Optisulin is contraindicated in the case of hypoglycemia. It should never be mixed with other products.

Further Reading

http://www.aventis.com
http://www.eudra.org
http://www.fda.gov
http://www.optisulin.com
McKeage, K. and Goa, K.L., Insulin glargine: a review of its therapeutic use as a long-acting agent for the management of type 1 and 2 diabetes mellitus, *Drugs*, 61, 1599–1624, 2001.

Orthoclone OKT-3

Product Name: Orthoclone OKT-3 (trade name)

Murumonab-CD3 (common name)

Description: Orthoclone OKT-3 is an intact murine monoclonal antibody produced by classical hybridoma technology. The antibody binds specifically to a 20-kDa surface glycoprotein found exclusively on the surface of mature T lymphocytes and medullary thymocytes. This surface glycoprotein is also known as the CD-3 receptor, and binding of antibody hinders cellular functioning by interfering with cell-mediated immune responses. Particularly affected is the functioning of cytotoxic T lymphocytes that are involved in the recognition and destruction of foreign antigens, such as those involved in mediating acute allograft rejection. Orthoclone OKT-3, which was the first monoclonal antibody to be approved for *in vivo* use, is generally supplied in packages of five ampules. Each 5-ml ampule contains 5 mg of active ingredient.

Approval Date: 1996 (U.S.)

Therapeutic Indications: Orthoclone OKT-3 was originally indicated for the treatment of acute allograft rejection in renal transplant patients. It is now also approved for the treatment of steroid-resistant acute allograft rejection in cardiac and hepatic transplant patients. The dosage of other immunosuppressive agents used in conjunction with Orthoclone OKT-3 should be reduced to the lowest level compatible with an effective therapeutic response.

Manufacturer: Ortho Biotech Products, LP, Raritan, NJ 08869, http://www.orthobiotech.com

Marketing Ortho Biotech Products, LP, Raritan, NJ 08869, http://www.orthobiotech.com

Manufacturing

Orhtoclone OKT-3 is manufactured by classical hybridoma technology. The antibody-producing hybridoma cell was originally developed by immunizing a mouse with human peripheral T lymphocytes bearing the cell surface CD-3 antigen. Antibody-producing B cells extracted from the animal's spleen were then incubated with murine myeloma cells in the presence of polyethylene glycol, thereby promoting cellular fusion. This was followed by screening studies in order to identify a hybridoma cell line producing the appropriate anti-CD-3 monoclonal antibody. This primary hybridoma cell line was shown to be free of murine viruses and was used to generate a cell bank system.

Production of a batch of Orthoclone OKT-3 is initiated by injection of hybridoma cells into the peritoneal cavity of male CAF-3 mice. The hybridoma cells grow and divide in the peritoneal cavity and secrete the anti CD-3 antibody into the cavity. The resultant antibody-containing ascitic fluid is harvested from the mice and clarified via centrifugation. This crude preparation may be stored frozen. Downstream purification of the antibody relies primarily upon an ammonium sulphate precipitation step, followed by anion exchange chromatography. After purification, the excipients (sodium chloride, polysorbate 80, and sodium phosphate buffer constituents) are added. After potency adjustment and sterile filtration, the product, which displays a pH of 6.5 to 7.5, is aseptically filled into the final product ampules.

Overview of Therapeutic Properties

Orthoclone OKT-3 reverses graft rejection by blocking the function the T lymphocytes that play a central role in acute allograft rejection. *In vitro* studies indicate that the product blocks both the generation and functioning of immune effector cells. Binding results in an early activation of T cells, which leads to cytokine release. This is followed by blocking of the T-cell function. *In vivo*, Orthoclone OKT-3 has been shown to react with both peripheral blood T lymphocytes and T lymphocytes residing in body tissues. Clinical trials indicate its effectiveness in the prevention of acute renal, cardiac, and hepatic allograft rejection.

The recommended dosage regime for adults entails daily administration of 5 mg of product by single-dose intravenous injection over a period of not more than 1 minute. The initial recommended dose in pediatric patients is 2.5 mg/day if they weigh less than 30 kg. A course of treatment is usually 10 to 14 days. In the case of acute renal rejection, treatment should begin upon diagnosis. In the case of steroid-resistant cardiac or hepatic allograft rejection, treatment should begin when it is obvious that rejection has not been reversed by an appropriate course of corticosteroid therapy.

Orthoclone OKT-3 is contraindicated in patients who are hypersensitive to the product, or any other product of murine origin, who display heart failure or fluid overload, who are pregnant or breast feeding, who display uncontrolled hypertension, or who have a history of seizures or are predisposed to seizures. Anaphylactic and anaphylactoid reactions may occur following administration of OKT-3. Serious or fatal systemic cardiovascular and central nervous systems reactions have also been reported. Pretreatment with methylprednisolone is recommended to minimize symptoms of cytokine release syndrome.

Further Reading

Ettenger, R.B. et al., OKT-3 for rejection reversal in paediatric renal transplantation, *Clin. Transplant.*, 2, 180–184, 1988.

Gaston, R.S. et al., OKT-3 first dose reaction: association with T cell subsets and cytokine release, *Kidney Int.*, 39, 141-148, 1991.

Schroeder, T.J. et al., Immunological monitoring during and following OKT-3 therapy in children, *Clin. Transplant.*, 5, 191–196, 1991.

Osigraft/OP-1 Implant

Product Name:	Osigraft (trade name E.U.) OP-1 Implant (trade name U.S.) Eptotermin alfa (proposed international nonproprietary name)
Description:	OP-1 (osteogenic protein-1; also known as bone morphogenetic protein-7) is a recombinant human osteogenic protein that stimulates bone formation. It is produced in Chinese hamster ovary (CHO) cells by recombinant DNA technology. The product is presented as a lyophilized powder to be resuspended with the provided solvent for implantation at the bone site via surgical procedure. Each vial contains 3.5 mg OP-1 and 1 g collagen
Approval Date:	2001 (E.U. and U.S.)
Therapeutic Indications:	Osteogenic protein-1:BMP-7 is indicated in skeletally mature patients for the treatment of trauma-related fractures of the tibia that do not heal after 9 months of treatment and in cases in which autografts are not feasible or have failed.
Manufacturer:	Howmedica International S. de R.L., Division of Stryker Corporation, Raheen Industrial Estate, Raheen, Limerick, Ireland (manufacturer is responsible for batch import and release in the European Economic Area) Stryker Biotech, 35 South Street, Hopkinton, MA 01748, http://www.strykercorp.com (U.S.)
Marketing:	Howmedica International S. de R.L., Division of Stryker Corporation, Raheen Industrial Estate, Raheen, Limerick, Ireland (E.U.) Stryker Biotech, 35 South Street, Hopkinton, MA 01748, http://www.strykercorp.com (U.S.)

Manufacturing

Osteogenic protein-1:BMP-7 is a recombinant human osteogenic protein. It is produced by recombinant DNA technology in a CHO cell line. The gene encoding the protein was amplified from a human cDNA library by using a probe based on bovine osteogenic protein sequences. The protein is expressed and purified from the culture via several chromatographic and filtration steps. The purified product is then mixed with its collagen carrier, a type 1 collagen chemically extracted from bovine bones of U.S. origin. The final product is provided in a lyophilized form, with a separate solvent containing sodium chloride.

The shelf life of the product is 12 months when stored at 2 to 8°C. The quality and safety of the product are ensured by extensive tests.

Overview of Therapeutic Properties

OP-1 is used to stimulate the formation of new bone in fractures of the tibia that failed to heal within 9 months of treatment, or in cases where autografts were unsuccessful. It is applied at the site of fracture during surgery in skeletally mature patients. The active component, osteogenic protein-1, stimulates regeneration of the bone via recruitment of mesenchymal stem cells, which undergo differentiation while the collagen scaffold provides support for the new bone cells. The use of OP-1 was found to exhibit the same efficacy as autografts in regard to reducing pain and reducing weight-bearing effects, but autoradiographic analysis indicated better healing in the case of autografts.

The components of OP-1, osteogenic protein-1 and bovine collagen, elicited an immunoresponse in 14% of patients treated, although this response diminished over time and did not appear to influence the outcome of the treatment. OP-1 is contraindicated during pregnancy and lactation and in the case of autoimmune diseases.

Further Reading

http://www.eudra.org
http://www.fda.gov
http://www.op1.com
http://www.strykercorp.com
Friedlaender, G.E. et al., Osteogenic protein-1 (bone morphogenetic protein-7) in the treatment of tibial nonunions, *J. Bone Joint Surg. Am.*, 83(A Suppl. 1), 151–158, 2001.

Ovitrelle/Ovidrel

Product Name: Ovitrelle/Ovidrel (trade name)

Choriogonadotropin alfa (international nonproprietary name)

Description: Ovitrelle (Ovidrel in the U.S.) is a recombinant human chorionic gonadotropin (hCG). The hormone is a heterodimeric glycoprotein, consisting of a 92-amino acid α subunit and a 145-amino acid β subunit. It is produced by recombinant DNA technology in Chinese hamster ovary (CHO) cells and is provided as a lyophilized powder (250 μg/vial) with a solvent for reconstitution prior to subcutaneous administration.

Approval Date: 2001 (E.U. and U.S.)

Therapeutic Indications: Ovitrelle is indicated for the treatment of female infertility due to anovulation and for patients undergoing assisted reproductive techniques. Ovitrelle triggers final follicle maturation and luteinization after follicle stimulation.

Manufacturer: Industria Farmaceutica Serono S.p.A., Zona Industriale di Modugno, 70123 Bari, Italy, http://www.serono.com

Marketing: Serono Europe Limited, 56 Marsh Wall, London E14 9TP, U.K. (E.U.)
Serono Inc., One Technology Place, Rockland MA, 02370, http://www.seronousa.com (U.S.)

Manufacturing

Ovitrelle is a recombinant hCG produced in a CHO cell line by using recombinant DNA technology. Downstream processing entails an ultrafiltration step, several chromatographic steps, and a nanofiltration step, which aims to remove any viral particles that may be present. The final product, which is very similar to the natural molecule in its activity and glycosylation pattern, is provided as a lyophilized powder. Ovitrelle contains the active substance choriogonadotropin alfa and the following excipients: sucrose (as a bulking agent), phosphoric acid, and sodium hydroxide.

The shelf life of the lyophilized product is 24 months when stored at a temperature lower than 25°C.

Overview of Therapeutic Properties

Human chorionic gonadotropin is used in infertility treatments to trigger final maturation of follicles and luteinization after follicle stimulation has been induced. Ovitrelle is the first recombinant hCG hormone and replaces the hormone of human origin that was previously extracted from the urine of pregnant or postmenopausal women. Ovitrelle has an activity almost identical to the natural hormone and offers the advantage of being more highly purified. Ovitrelle is used in the case of anovulation and in women undergoing assisted reproductive techniques. It is administered subcutaneously 35 hours after optimal follicle stimulation has been achieved. Ovitrelle has been shown to have at least the same efficacy as the hormone extracted from urine, with the same number of oocytes retrieved after follicle stimulation.

Reactions at the site of injection, extrauterine pregnancies, multiple gestations, ovarian hyperstimulation, and general gastrointestinal disorders were the most commonly reported side effects. Ovitrelle is contraindicated during pregnancy and lactation.

Further Reading

http://www.eudra.org
http://www.fda.gov
http://www.serono.com

Chang, P. et al., Recombinant human chorionic gonadotropin (rhCG) in assisted reproductive technology: results of a clinical trial comparing two doses of rhCG (Ovidrel) to urinary hCG (Profasi) for induction of final follicular maturation in *in vitro* fertilization-embryo transfer, *Fertil. Steril.*, 76, 67–74, 2001.

Driscoll, G.L. et al., A prospective, randomized, controlled, double-blind, double-dummy comparison of recombinant and urinary hCG for inducing oocyte maturation and follicular luteinization in ovarian stimulation, *Hum. Reprod.*, 15, 1305–1310, 2000.

Pegasys

Product Name:	Pegasys (trade name) Peginterferon alfa-2a (international nonproprietary name)
Description:	Pegasys is a modified pegylated form of Roferon A, a recombinant human interferon alfa-2a. The 20-kDa interferon moiety is produced in *Escherichia coli* and a single 40-kDa bis-monomethoxy polyethylene glycol (PEG) is subsequently covalently attached to a lysine residue in the interferon's backbone. The 60-kDa product is supplied as a solution (180 μg/ml) for subcutaneous administration.
Approval Date:	2002 (U.S.)
Therapeutic Indications:	Pegasys is indicated for the treatment of adult patients with chronic hepatitis C as a monotherapy or in combination with ribavirin (Copegus).
Manufacturer:	Hoffman-La Roche, Inc., 340 Kingsland Street, Nutley, NJ 07110-1199, http://www.rocheusa.com
Marketing:	Hoffman-La Roche, Inc., 340 Kingsland Street, Nutley, NJ 07110-1199, http://www.rocheusa.com

Manufacturing

Pegasys is a modified form of the recombinant human interferon alfa-2a, Roferon A. It differs from the natural molecule in having an additional methionine residue at its N-terminus and in lacking glycosylation. It also has an additional single-branched bis-monomethoxy PEG chain, linked *in vitro* to a lysine residue via an amide bond, which is not present in Roferon A. It is produced in *E. coli* and undergoes several purification steps, including affinity, ion exchange, and size exclusion chromatography. The final product is presented as a solution containing peginterferon alfa-2a (active), as well as sodium chloride, polysorbate 80, benzyl alcohol, sodium acetate trihydrate, and acetic acid as excipients. Pegasys is supplied in vials for subcutaneous administration.

Overview of Therapeutic Properties

Pegasys exhibits the classical biological activities of type 1 interferons, such as inhibition of viral replication, mediation of cell proliferation, and immunomodulation. It is indicated in the treatment of adult patients with chronic hepatitis C who have compensated liver disease and have not been previously treated with interferon alfa. Pegasys retains the activity of its nonpegylated form, Roferon A, and has the advantage of requiring only one weekly administration, as opposed to three in the case of Roferon A. This is due to the presence of the PEG side chain in Pegasys, which increases the *in vivo* half-life of the molecule and reduces its immunogenecity. Pegasys is indicated as a monotherapy or in combination with ribavirin (Copegus). Clinical trials showed an increased response in patients receiving Pegasys subcutaneously once a week for 48 weeks, compared with patients receiving Roferon A three times a week over the same period. The combination therapy of Pegasys administered with Copegus showed a further increase in the response. A lower response was observed in patients with poor prognostic factors, including those who were more than 40 years old, overweight, or had cirrhosis.

The most commonly reported side effects were flu-like symptoms, headaches, and skin and gastrointestinal disorders. Severe infections, pancreatitis, and cardiovascular, pulmonary, and ophthalmologic disorders were also observed, as well as depression and suicidal behavior. Pegasys should not be used in combination with Didanosine, Stavudine, or Zidovudine. It is contraindicated during pregnancy and lactation, in men whose female partners are pregnant, and in patients with autoimmune disorders.

Further Reading

http://www.fda.gov
http://www.pegasys.com
http://www.rocheusa.com

Hadziyannis, S.J. and Papatheodoridis, G.V., Peginterferon-alpha (2a) (40 kDa) for chronic hepatitis C, *Expert Opin. Pharmacother.*, 4, 541–551, 2003.

Keating, G.M. and Curran, M.P., Peginterferon-alpha-2a (40kD) plus ribavirin: a review of its use in the management of chronic hepatitis C, *Drugs*, 63, 701–730, 2003.

Rajender, R. K. et al., Use of peginterferon alfa-2a (40 KD) (Pegasys) for the treatment of hepatitis C, *Adv. Drug Deliv. Rev.*, 54, 571–586, 2002.

PegIntron

Product Name:	PegIntron (trade name) Peginterferon alfa-2b (international nonproprietary name)
Description:	PegIntron is a recombinant modified human interferon alfa-2b. It is derived from the product Intron A. It differs from the natural molecule by the addition of a polyethylene glycol polymer strand conjugated to the molecule. It is apparently identical to the product marketed with the trade name ViraferonPeg. PegIntron is produced by recombinant DNA technology in *Escherichia coli* cells and is presented as a powder to be resuspended before subcutaneous injection. It is also provided in vials and cartridges to be used with multidose devices.
Approval Date:	2000 (E.U.); 2001 (U.S.)
Therapeutic Indications:	PegIntron is indicated for the treatment of adult patients with chronic hepatitis C.
Manufacturer:	SP (Brinny) Company, Innishannon, County Cork, Ireland, http://www.schering-plough.com (manufacturer is responsible for batch release in the European Economic area) Schering Corporation, Galloping Hill Road, Kenilworth, NJ 07033, http://www.schering.com (U.S.)
Marketing:	SP Europe, 73 rue de Stalle, 1180 Bruxelles, Belgium, http://www.schering-plough.com (E.U.) Schering Corporation, Galloping Hill Road, Kenilworth, NJ 07033, http://www.schering.com (U.S.)

Manufacturing

PegIntron is a modified recombinant human interferon alfa-2b. It is produced by recombinant DNA technology in *E. coli* after transformation of the plasmid encoding the interferon gene from human leukocytes. After purification interferon alfa-2b is covalently linked to a methoxy polyethylene glycol. The final product contains the active substance, consisting of isomeric forms of peginterferon alfa-2b, and the following excipients: dibasic sodium phosphate, monobasic sodium phosphate, sucrose, and polysorbate 80. The product is provided in a lyophilized form with the solvent for reconstitution prior to subcutaneous injection. Multidose devices are also available.

The shelf life of the product is 24 to 36 months when stored at 2 to 8°C. Extensive characterization of the product has been carried out, and tests that ensure the product's quality and safety include SDS-PAGE, mass spectroscopy, and chromatographic procedures.

Overview of Therapeutic Properties

PegIntron is a modified form of the product Intron A, the human interferon alfa-2b, used to treat a variety of conditions, including infectious diseases. PegIntron differs from Intron A by the presence of a covalently attached polyethylene glycol molecule. This modification results in a longer half-life of the molecule but unaltered activity: activation of molecules that inhibit viral replication in infected cells, induction of phagocytosis in macrophages, and cytotoxicity of lymphocytes and suppression of cell proliferation.

PegIntron is indicated in the treatment of adult patients with chronic hepatitis C when there is evidence of viral replication. A longer half-life of PegIntron results in a lower required dosage compared to Intron A, with the associated advantage of reduced side effects. PegIntron should be administered once a week in combination with ribavirin. Administration should be subcutaneous, with the dosage adjusted for individual needs, and should continue for at least 6 months and up to 1 year.

PegIntron, when administered in combination with ribavirin, resulted in a better response than Intron A. The most common side effects were similar to those reported for Intron A and included flu-like symptoms, loss of appetite, nausea, variation in hematic and enzymatic values, depression, and visual disorders. Suicidal behavior and cardiac disorders were also reported, though rarely. PegIntron is contraindicated during pregnancy and lactation.

Further Reading

http://www.eudra.org
http://www.fda.gov
http://www.pegintron.com
http://www.schering-plough.com

Manns, M.P. et al., Peginterferon alfa-2b plus ribavirin compared with interferon alfa-2b plus ribavirin for initial treatment of chronic hepatitis C: a randomised trial, *Lancet*, 358, 958–965, 2001.

Patel, K. and McHutchison, J., Peginterferon alpha-2b: a new approach to improving response in hepatitis C patients, *Expert Opin. Pharmacother.*, 2, 1307–1315, 2001.

Primavax (withdrawn from market)

Product Name:	Primavax (trade name) Diphtheria, tetanus, and hepatitis B vaccine (international nonproprietary name)
Description:	Primavax was a trivalent vaccine containing diphtheria toxoid, tetanus toxoid, and recombinant hepatitis B surface antigen (rHBsAg). Each monodose preparation contained 30 IU purified adsorbed diphtheria toxoid, 40 IU tetanus toxoid, and 5 μg rHBsAg in a final volume of 0.5 ml. The product was presented in a prefilled glass syringe as a suspension for intramuscular administration.
Approval Date:	1998 (E.U.)
Withdrawal Date:	2000
Therapeutic Indications:	Primavax was indicated for active immunization against hepatitis B (caused by all known subtypes), diphtheria, and tetanus in infants for primary vaccination and booster shots.
Manufacturer:	Pasteur Merieux MSD, 8 rue Jonas Salk, 69367 Lyon Cedex 07, France
Marketing:	Pasteur Merieux MSD, 8 rue Jonas Salk, 69367 Lyon Cedex 07, France

Manufacturing

Hepatitis B surface antigen was produced by recombinant DNA technology in an engineered strain of *Saccharomyces cerevisiae*. The product was purified by a combination of high resolution chromatogaphic steps. Details of manufacture are not freely available since the product was withdrawn from the market.

Overview of Therapeutic Properties

Clinical trials proved the product to be safe and effective in eliciting protective immunity against diphtheria, tetanus, and hepatitis B in infants. Details of therapeutic and related properties are not freely available since the product was withdrawn from the market.

PROCOMVAX/COMVAX

Product Name: PROCOMVAX (trade name E.U.)

COMVAX (trade name U.S.)

Hemophilus B conjugate (meningococcal protein conjugate) and hepatitis B (recombinant) vaccine (common name)

Description: PROCOMVAX (or COMVAX in the U.S.) is a vaccine consisting of antigenic components present in other previously marketed vaccines. PROCOMVAX contains polyribosylribitol phosphate from *Haemophilus influenzae* type b coupled to the outer membrane protein complex of *Neisseria meningitidis* and the major surface antigen of the hepatitis B virus produced by recombinant DNA technology in *Saccharomyces cerevisiae*. It is presented as a suspension for intramuscular injection. Each single-dose vial contains 7.5 μg of Hemophilus b polyribosylribitol phosphate (PRP) conjugated to 125 μg of the outer membrane protein complex of *N. meningitidis* and 5.0 μg of HBsAg as active substance, in a total volume of 0.5 ml.

Approval Date: 1999 (E.U.); 1996 (U.S.)

Therapeutic Indications: PROCOMVAX is indicated for active immunization against *H. influenzae* type b and hepatitis B in infants and children from 6 weeks to 15 months of age.

Manufacturer: Merck Sharp and Dohme, B.V., Waarderweg 39, P.O. Box 581, 2003 PC Haarlem, the Netherlands, http://www.merck.com (manufacturer is responsible for batch import and release in the European Economic Area)

Merck and Co., Inc., P.O. Box 4, Sumneytown Pike, West Point, PA 19486, http://www.merck.com (U.S.)

Marketing: Aventis Pasteur MSD SNC, 8 rue Jonas Salk, 69007 Lyon, France, http://www.aventispasteur.com (E.U.)

Merck and Co., Inc., P.O. Box 4, Sumneytown Pike, West Point, PA 19486, http://www.merck.com (U.S.)

Manufacturing

PROCOMVAX is a combination of two vaccines previously marketed individually. It contains polyribosylribitol phosphate from *H. influenzae* type b coupled to the outer membrane protein complex from *N. meningitidis* as the antigenic component for *H. influenzae,* and the recombinant hepatitis B major surface antigen as the antigenic component for the hepatitis B virus. Polyribosylribitol phosphate is a capsular polysaccharide purified from a culture of *H. influenzae* type b strain. Purification involves phospholipase D treatment to separate polysaccharides from phospholipids, as well as several precipitation steps to remove contaminants. The outer membrane protein complex is isolated by extraction with detergent from a culture of *N. meningitidis,* following phenol inactivation. Polyribosylribitol phosphate and the outer membrane protein complex are derivitized for chemical coupling. The coupled compound is then further purified from contaminants. The major membrane protein, the S protein, of the hepatitis B virus is produced in a genetically modified *S. cerevisiae* strain using recombinant DNA technology. Purification includes several chromatographic steps and filtration procedures. The purified protein is inactivated by formaldehyde treatment and adsorbed on aluminum hydroxide, following which the two antigenic components are mixed into the vaccine formulation. The final product contains the antigenic proteins, *Haemophilus influenzae* type b PRP coupled to the outer membrane protein complex of *N. meningitidis,* and the recombinant major surface antigen from the hepatitis B virus. It also contains the following excipients: aluminium hydroxide (as an adjuvant), sodium borate, and sodium chloride. PROCOMVAX is presented as a suspension for intramuscular injection.

The shelf life of the product is 36 months when stored at 2 to 8°C. The quality and safety of the vaccine are ensured by extensive testing for sterility, stability, identity, potency, and the presence of contaminants.

Overview of Therapeutic Properties

PROCOMVAX is indicated for immunization of infants and children from 6 weeks to 15 months of age against *H. influenzae* type b and the hepatitis B virus. Three doses of PROCOMVAX should be administered intramuscularly at 2, 4, and 12 to 15 months of age, or with at least a 2-month interval between the first and second doses and an 8- to 11-month interval between the second and third doses.

PROCOMVAX proved to be effective in eliciting protection against *H. influenzae* type b and hepatitis B and could be administered simultaneously with the diphtheria, tetanus, and whole-cell pertussis vaccines (oral poliomyelitis vaccine, inactivated poliomyelitis vaccine), and the Merck measles, mumps, and rubella virus vaccine. PROCOMVAX exhibited side effects comparable to those commonly reported with other vaccines, such as reactions at the sites of injection and flu-like symptoms. PROCOMVAX should not be administered in cases of fever.

Further Reading

http://www.aventispasteur.com
http://www.eudra.org
http://www.fda.gov
http://www.merck.com
Petersen, K. et al., Immunogenicity of a combined hepatitis B and Haemophilus influenzae type b conjugate vaccine in Alaska Native infants, *Int. J. Circumpolar. Health,* 57(Suppl. 1), 285–292, 1998.
West, D.J. et al., Safety and immunogenicity of a bivalent Haemophilus influenzae type b/hepatitis B vaccine in healthy infants, *Pediatr. Infect. Dis. J.*, 16, 593–599, 1997.

Proleukin

Product Name: Proleukin (trade name)

Aldesleukin (international nonproprietary name)

Description: Aldesleukin is a modified form of human interleukin-2 (IL-2). It is produced in *Escherichia coli* using recombinant DNA technology and differs slightly from the natural human molecule: it lacks a serine residue at the N terminus, it has a serine residue replacing a cysteine at position 125, it is unglycosylated, and its aggregation state is different from that of the natural molecule. Nevertheless, it retains the biological activity of the natural molecule. Proleukin is provided as a lyophilized powder to be resuspended with water for injection to a proleukin concentration of 1.1 mg/ml before intravenous administration.

Approval Date: 1992, for metastatic renal cell carcinoma; 1998, for metastatic melanoma (U.S.)

Therapeutic Indications: Proleukin is indicated for the treatment of metastatic renal cell carcinoma and metastatic melanoma in adult patients.

Manufacturer: Chiron Corporation, 4560 Horton Street, Emeryville, CA 94608-2916, http://www.chiron.com

Marketing: Chiron Corporation, 4560 Horton Street, Emeryville, CA 94608-2916, http://www.chiron.com

Manufacturing

Aldesleukin is a modified form of human IL-2 produced in *E. coli* using recombinant DNA technologies. The hydrophobic protein is extracted from the culture medium into a solution containing sodium dodecyl sulfate, whereupon it is subjected to chemical reduction, size exclusion chromatography, oxidation, reverse phase HPLC, and further chromatographic and filtration purification procedures. The purified final product is provided as a lyophilized powder containing aldesleukin (active), as well as mannitol, sodium dodecyl sulphate, monobasic sodium phosphate, and dibasic sodium phosphate as excipients. Proleukin is resuspended before intravenous administration. Once resuspended, it forms a microaggragate corresponding to the size of 27 molecules of aldesleukin.

The shelf life of the product is 18 months when stored at 2 to 8°C. Extensive tests are undertaken to ensure the quality and safety of the product. These include amino acid sequencing, tryptic peptide map analysis, reducing and non-reducing SDS-PAGE, RP-HPLC, isoelectric focusing, lymphocyte proliferation bioassay, and tests for sterility, bacterial, and viral safety.

Overview of Therapeutic Properties

Aldesleukin is a modified form of the human IL-2. It is produced *in vivo* by activated T cells and regulates the immune system by binding to a specific receptor on various immune cells. In this way, it induces activation and proliferation of T cells, natural killer cells, lymphokine activated killer cells, and other lymphocytes, as well as the production of cytokines such as IL-1, gamma interferon, and tumor necrosis factor. Aldesleukin retains the activity of natural IL-2. However, due to lack of glycosylation (the glycocomponent promotes elimination of the natural molecule), it exhibits an extended half-life.

Proleukin is indicated for the treatment of adult patients with metastatic renal cell carcinoma and metastatic melanoma. Proleukin is administered as a 15-minute intravenous infusion every 8 hours for 14 doses and is repeated for another 14 doses after a rest of 9 days. The treatment should be repeated 7 weeks after hospital discharge only in patients who experienced some tumor shrinkage in the previous treatment. In studies of metastatic renal cell carcinoma, a complete response was obtained in 7% of patients and a partial response in 8%, while in metastatic melanoma studies a complete response was obtained in 6% and partial response in 10% of patients. Most of the patients did not receive the whole 28-dose regimen, due to suspension of the treatment upon observation of side effects.

Adverse side effects were reported with a high frequency and severity, but most were reversible after discontinuation of the treatment. Capillary leak syndrome was commonly reported. In studies of patients with renal cell carcinoma, the rate of drug-related death was 4%, and in patients with metastatic melanoma the rate was 2%. More than 65% of the patients developed non-neutralizing antibodies against proleukin. Proleukin is restricted to patients with normal cardiac and pulmonary functions and is contraindicated during pregnancy and lactation.

Further Reading

http://www.chiron.com
http://www.fda.gov
http://www.proleukin.com
Atkins, M.B. et al., High-dose recombinant interleukin-2 therapy for patients with metastatic melanoma: analysis of 270 patients treated between 1985 and 1993, *J. Clin. Oncol.*, 17, 2105–2116, 1999.
Baigent, G., Recombinant interleukin-2 (rIL-2), aldesleukin, *J. Biotechnol.*, 95, 277–280, 2002.
Fyfe, G. et al., Results of treatment of 255 patients with metastatic renal cell carcinoma who received high-dose recombinant interleukin-2 therapy, *J. Clin. Oncol.*, 13, 688–696, 1995.
Shanafelt, A.B. et al., A T-cell-selective interleukin-2 mutein exhibits potent antitumor activity and is well tolerated *in vivo*, *Nat. Biotechnol.*, 18, 1197–1202, 2000.

ProstaScint

Product Name: ProstaScint (trade name)

Capromab pendetide (international nonproprietary name)

Description: ProstaScint consists of a murine IgG1 kappa monoclonal antibody conjugated to a tripeptide linker-chelator in order to be coupled to the radioisotope Indium 111 (In 111) before use. The monoclonal antibody is produced in a hybridoma cell line and is directed against the prostate specific membrane antigen, a glycoprotein expressed on the surface of prostate epithelial cells, including malignant cells. ProstaScint is provided as a solution as part of a kit for the reconstitution of the radiolabeled molecule; ProstaScint is coupled to In 111, which is not provided with the kit, before intravenous administration for diagnostic imaging.

Approval Date: 1996 (U.S.)

Therapeutic Indications: ProstaScint, when coupled to In 111, is used in diagnostic imaging in newly diagnosed patients with clinically localized biopsy-proven prostate cancer who are at high risk for pelvic lymphonode metastasis, and in post-prostatectomy patients in whom there is a high suspicion of occult metastatic disease.

Manufacturer: Cytogen Corporation, 600 College Road East, Princeton, NJ 08540, http://www.cytogen.com

Marketing: Cytogen Corporation, 600 College Road East, Princeton, NJ 08540, http://www.cytogen.com

Manufacturing

The monoclonal antibody CYT-351, also known as 7E11-C5, binds the prostate-specific membrane antigen, which is expressed on the surface of normal and malignant prostate epithelial cells. It is produced in a bioreactor by using a hybridoma cell line obtained by fusion of a myeloma cell line with spleen cells from mice immunized with membrane extracts and whole cells of a human prostate adenocarcinoma. The antibody is purified using several chromatographic steps, including gel filtration, anion exchange, cation exchange and affinity chromatography, and filtration procedures. The purified antibody is then conjugated to the linker-chelator glycyl-tyrosyl-(N-ε-diethylenetriaminepentaacetic acid)-lysine hydrochloride by oxidation with sodium metaperiodate and is referred to as CYT-356. The final product is presented as a solution containing capromab pendetide, or CYT-356, sodium phosphate, and sodium chloride. ProstaScint is presented in a kit with a second vial containing a solution of sodium acetate used to buffer In 111 before labeling of the antibody.

ProstaScint has a shelf life of 24 months when stored at 2 to 8°C. Extensive tests carried out to verify the quality and safety of the product include N-terminal amino acid sequencing, SDS-PAGE, isoelectric focusing, protein A assay, immunoreactivity assays, and tests to ensure microbiological and viral safety such as mycoplasma assay, S^+L^-focus, XC plaque, and reverse transcriptase testing.

Overview of Therapeutic Properties

ProstaScint is indicated for diagnostic radioimaging in patients with biopsy-proven prostate cancer who are at high risk of metastasis. The antibody component of ProstaScint is directed against the prostate-specific membrane antigen that is expressed on the surface of prostate epithelial cells. ProstaScint reacts primarily with metastatic and primary prostatic carcinoma and, to a lesser degree, with benign prostatic hypertrophy and normal prostate tissues. No reactivity was found with other malignant cell lines. A cross-reactivity was observed with normal skeletal muscle, heart muscle, kidney proximal tubule, and skin. ProstaScint should always be used in combination with other diagnostic tests due to the high rate of false-positive and false-negative results.

ProstaScint is administered intravenously after coupling to the radioisotope In 111. A first SPECT (single photon emission computed tomography) imaging is performed 30 minutes after the administration of the radiolabeled product to obtain a background image of the pelvis, and a second SPECT is performed 72 to 120 hours later.

With an accuracy of 55 to 70%, ProstaScint was found to improve the diagnosis of metastatic and primary prostatic carcinoma when used in combination with other diagnostic tools. Minor side effects, such as injection site reactions, hypotension, hypertension, and elevated liver enzymes, were observed after the use of ProstaScint in 4% of the patients. High levels of human antimouse antibodies were detected in a small number of patients.

Further Reading

http://www.cytogen.com
http://www.fda.gov
http://www.prostascint.com
Freeman, L.M. et al., The role of (111)In Capromab Pendetide (Prosta-ScintR) immunoscintigraphy in the management of prostate cancer, *Q. J. Nucl. Med.*, 46, 131–137, 2002.
Petronis, J.D. et al., Indium-111 capromab pendetide (ProstaScint) imaging to detect recurrent and metastatic prostate cancer, *Clin. Nucl. Med.*, 23, 672–677, 1998.
Raj, G.V. et al., Clinical utility of indium 111-capromab pendetide immunoscintigraphy in the detection of early, recurrent prostate carcinoma after radical prostatectomy, *Cancer*, 94, 987–996, 2002.
Sodee, D.B. et al., Multicenter ProstaScint imaging findings in 2154 patients with prostate cancer, *Urology*, 56, 988–993, 2000.

Protropin

Product Name: Protropin (trade name)

Somatrem (international nonproprietary name)

Description: Protropin is a 22-kDa, 192-amino acid recombinant human growth hormone that differs from the natural hormone in having an additional methionine residue at the N terminus. Protropin completely retains the growth-inducing activity of the natural molecule. Protropin is produced by recombinant DNA technology in *Escherichia coli* and is presented in a lyophilized form (in vials containing either 5 or 10 mg active), to be resuspended with the provided solvent (water for injection containing 0.9% benzyl alcohol) before intramuscular or subcutaneous administration.

Approval Date: 1985 (U.S.)

Therapeutic Indications: Protropin is indicated for long-term treatment of children with growth failure due to insufficient endogenous growth hormone.

Manufacturer: Genentech, Inc., 1 DNA Way, South San Francisco, CA 94080-4990, http://www.gene.com

Marketing: Genentech, Inc., 1 DNA Way, South San Francisco, CA 94080-4990, hhttp://www.gene.com

Manufacturing

Protropin is a recombinant human growth hormone that differs from the natural molecule by containing an additional methionine residue at the N-terminus of the protein. Protropin is produced by recombinant DNA technology in a modified *E. coli* strain. The molecule is extensively purified using multiple chromatographic steps, and the final product is presented in a lyophilized form to be resuspended with the provided solvent prior to intramuscular or subcutaneous administration. Protropin contains the recombinant growth hormone somatrem (active), as well as mannitol, monobasic sodium phosphate, and dibasic sodium phosphate as excipients. The solvent provided with the lyophilized product contains benzyl alcohol as a preservative.

Overview of Therapeutic Properties

Protropin is a recombinant growth hormone. It retains the activity of the natural molecule, which consists of stimulating skeletal growth directly and by increasing levels of insulin-like growth factor-I. Protropin is indicated for long-term treatment of children who suffer from growth failure due to inadequate levels of endogenous growth hormone. Protropin should be administered daily via intramuscular or subcutaneous injection. It proved to be effective in stimulating growth, leading to increases in height.

Protropin was well tolerated with only mild side effects, which included pain at the injection site and intracranial hypertension. Antibodies against the hormone, which lead to an attenuation of the effectiveness of the product, rarely developed. Protropin is contraindicated in patients with closed epiphyses, active neoplasia, acute respiratory failure, and illness due to complications after open heart surgery, abdomen surgery, or multiple accidental trauma.

Further Reading

http://www.fda.gov

http://www.gene.com

Blethen, S.L. et al., Factors predicting the response to growth hormone (GH) therapy in prepubertal children with GH deficiency, *J. Clin. Endocrinol. Metab.*, 76, 574–579, 1993.

Chen, S.A. et al., Bioequivalence of two recombinant human growth hormones in healthy male volunteers after subcutaneous administration, *Am. J. Ther.*, 2, 190–195, 1995.

Sherman, B.M., A national cooperative growth study of Protropin, *Acta Paediatr. Scand. Suppl.*, 337, 106–108, 1987.

Pulmozyme

Product Name: Pulmozyme (trade name)

Dornase alfa (international nonproprietary name)

Description: Dornase alfa is a 37-kDa, 260-amino acid recombinant human deoxyribonuclease I (DNase I), an endonuclease enzyme that selectively cleaves DNA. Dornase alfa is produced by recombinant DNA technology in Chinese hamster ovary (CHO) cells and is presented as a solution (1 mg active/ml) to be administered by inhalation with a nebulizer system.

Approval Date: 1993 (U.S.)

Therapeutic Indications: Pulmozyme is indicated in cystic fibrosis patients, in addition to standard therapies, to improve pulmonary function and to reduce the risk of infections requiring antibiotic treatment in patients with a forced vital capacity (greater than 40%).

Manufacturer: Genentech, Inc., 1 DNA Way, South San Francisco, CA 94080-4990, http://www.gene.com

Marketing: Genentech, Inc., 1 DNA Way, South San Francisco, CA 94080-4990, http://www.gene.com

Manufacturing

Dornase alfa is a recombinant human DNase I, an endonuclease similar to the natural molecule. The human gene has been cloned and expressed in a CHO cell line and is purified from the culture medium using multiple chromatographic and filtration procedures. The antibiotic gentamicin is used during fermentation, but no traces are detectable in the final, purified product. The final product is presented after extensive characterization as a solution to be administered for inhalation using an aerosol nebulizer device. Pulmozyme contains the active substance dornase alfa, as well as dihydrate calcium chloride and sodium chloride as excipients.

The quality and safety of the product are ensured by the use of numerous tests carried out throughout production and on the final product.

Overview of Therapeutic Properties

Cystic fibrosis is a common genetic disease with life-threatening implications. It results from the presence of two abnormal copies of the cystic fibrosis transmembrane conductance regulatory gene (CFTR), which normally regulates chloride channels to allow an electrochemical gradient from epithelial cells into the lumen, in a mechanism that is modulated by cAMP. When the CFTR gene product does not respond to cAMP modulation, the chloride channels remain closed, leading to decreased chloride secretion in the lumen and increased sodium reabsorpion, thus causing dehydration and thickening of secretions. Secretions also increase in viscosity due to the presence of polyanionic DNA, which is released after the disintegration of leukocytes due to infections.

Dornase alfa is an endonuclease that selectively degrades DNA by cleaving the phosphodiester linkage between polynucleotides, thus reducing the viscosity of secretions and resulting in improved pulmonary function. Cystic fibrosis also affects other organ systems, such as the digestive and reproductive systems, but pulmonary failure is the most common cause of death.

Pulmozyme should be administered once a day by aerosol inhalation using a compressed, air-driven nebulizer in addition to standard therapies for cystic fibrosis. Some patients require a second daily dose. Pulmozyme, which was developed for use in adult patients, is now available for children 5 years or older. The efficiency of Pulmozyme was demonstrated by an improvement in pulmonary function within a few days of the first administration and a reduction in infection within a few months of the beginning of treatment in patients with a forced vital capacity greater than 40%.

The most common side effects related to the use of Pulmozyme were voice alteration, rhinitis, pharyngitis, laryngitis, and rash. Neutralizing antibodies developed in 2% of the patients. Pulmozyme should be administered during pregnancy and lactation only if clearly necessary.

Further Reading

http://www.fda.gov
http://www.gene.com
Fuchs, H.J. et al., Effect of aerosolized recombinant human DNase on exacerbations of respiratory symptoms and on pulmonary function in patients with cystic fibrosis, *N. Engl. J. Med.*, 331, 637–642, 1994.

Robinson, P.J., Dornase alfa in early cystic fibrosis lung disease, *Pediatr. Pulmonol.*, 34, 237–241, 2002.

Wagener, J.S. et al., Aerosol delivery and safety of recombinant human deoxyribonuclease in young children with cystic fibrosis: a bronchoscopic study, *J. Pediatr.*, 133, 486–491, 1998.

Puregon/Follistim

Product Name: Puregon (trade name E.U.)

Follistim (trade name U.S.)

Follitropin beta (international nonproprietary name)

Description: Puregon (Follistim in the U.S.) is a recombinant human follicle-stimulating hormone (FSH). Like the native human molecule, puregon is a heterodimeric glycoprotein, consisting of a 92-amino acid α-subunit and a 111-amino acid β-subunit. Both are glycosylated and are held together by noncovalent interactions. Puregon is produced by recombinant DNA technology in Chinese hamster ovary (CHO) cells. It is provided for subcutaneous and intramuscular administration in different formulations: as a lyophilized form to be reconstituted before administration with the solvent provided, as a solution ready for injection, and as a solution for use in a multidose device.

Approval Date: 1996 (E.U.); 1997 (U.S.); 2002 (approval for treatment of male infertility, U.S.)

Therapeutic Indications: Puregon is indicated for treatment of female infertility and is used to stimulate follicle development in the ovary in the case of anovulation; it is also indicated for assisted reproductive technologies (ART) to induce ovulation. Puregon is also used to treat male infertility by stimulating spermatogenesis (tentative; U.S.).

Manufacturer: N.V. Organon, Kloosterstraat 6, Postbus 20, 5340 BH Oss, the Netherlands; Organon (Ireland) Ltd., P.O. Box 2857, Drynam Road, Swords, Co., Dublin, Ireland; and Organon S.A., Usine Saint Charles, 60590 Eragny sur Epte, France, http://www.organon.com (manufacturers is responsible for batch release in the European Economic Area)

Organon, Inc., 375 Mt. Pleasant Avenue, West Orange, NJ 07052, http://www.organoninc.com (U.S.)

Marketing: N.V. Organon, Kloosterstraat 6, Postbus 20, 5340 BH Oss, the Netherlands, http://www.organon.com (E.U.)
Organon, Inc., 375 Mt. Pleasant Avenue, West Orange, NJ 07052, http://www.organoninc.com (U.S.)

Manufacturing

Follitropin is produced by recombinant DNA technology in CHO cells, into which both the alfa and beta genes have been cloned. The recombinant product contains different isoforms with a very similar glycosylation pattern to the natural hormone. Follitropin beta is extensively purified from the culture medium using multiple chromatographic steps and is provided in different forms. A lyophilized form contains follitropin beta, sucrose, sodium citrate, polysorbate 20, sodium hydroxide, and hydrochloric acid, while a solvent containing sodium chloride is also provided for reconstitution before intramuscular or subcutaneous administration. Puregon is also available as a solution containing the active substance follitropin beta, as well as sucrose, sodium citrate, L-methionine, polysorbate 20, sodium hydroxide, and hydrochloric acid as excipients. The liquid formulation is also available with a multidose device, and in this case it also contains the antimicrobial agent benzyl alcohol.

The shelf life of the lyophilized product is 36 months when stored below 30°C. The liquid formulation has a shelf life of 24 months when stored at 2 to 8°C. The final product has been extensively analyzed using SDS-PAGE, Western blotting, ELISA, and hybridization assays to ensure its quality and safety.

Overview of Therapeutic Properties

Deficiency of the FSH leads to infertility. In these cases, the indicated treatment is hormone replacement therapy. The use of the recombinant hormone, follitropin beta, has replaced the use of FSH extracted from urine, with the advantage of not depending on large volumes of urine, as well as higher purity of the product. The recombinant follitropin beta is almost identical to the natural molecule, which is a combination of isoforms, but contains isoforms that are slightly more basic when compared to FSH from urine. The activity and specificity of the molecule are unchanged, resulting in a product with a higher potency than FSH from urine.

Puregon is used for the treatment of infertility in females and (tentatively) in males. In females, Puregon induces ovulation in women who do not respond to other ovulation treatment, such as clomiphene citrate, and it is used to induce multiple ovulations in women undergoing assisted reproductive techniques, such as *in vitro* fertilization. In males, Puregon is used to stimulate spermatogenesis.

Puregon is administered subcutaneously or intramuscularly at a dosage adjusted to suit individual needs. In the case of anovulation, daily administration is continued for 1 or 2 weeks, following which ovulation is induced by administration of human chorionic gonadotrophin (hCG). In the case of ART, the treatment is continued until follicles have reached the number and the size required, whereupon hCG is administered to induce ovulation and oocytes are retrieved after 35 hours. Three cycles of treatment are generally necessary.

In the case of male infertility, Puregon is usually administered in combination with hCG three times a week for 3 or 4 months, or up to 18 months as necessary. Puregon proved to be at least as effective as FSH from urine, while it exhibited a higher potency and can be administered at a lower dosage. Puregon was found to lead to a higher rate of pregnancies when used for ART and to lead to increased follicle development when administered with clomiphene citrate.

During treatment women need to be closely monitored in order to avoid over stimulation of the ovaries, which could lead to serious complications. Reactions at the injection sites, ectopic pregnancies, multiple gestations, and thromboembolism have been reported, as have breast development and acne in males. Puregon is contraindicated during pregnancy and lactation.

Further Reading

http://www.eudra.org
http://www.fda.gov
http://www.organon.com
http://www.puregon.com

Out, H.J., Follitropin beta (Puregon), in *Biopharmaceuticals, an Industrial Perspective*, Walsh, G. and Murphy, B., Eds., Dordrecht: Kluwer Academic Publisher, 125–148, 1999.

Out, H.J. et al., Recombinant follicle stimulating hormone (rFSH; Puregon) in assisted reproduction: more oocytes, more pregnancies. Results from five comparative studies, *Hum. Reprod. Update*, 2, 162–171, 1996.

Out, H.J. et al., A prospective, randomized, double-blind clinical trial to study the efficacy and efficiency of a fixed dose of recombinant follicle stimulating hormone (Puregon) in women undergoing ovarian stimulation, *Hum. Reprod.*, 14, 622–627, 1999.

Rapilysin/Retavase

Product Name:	Rapilysin (trade name E.U.) Retavase (trade name U.S.) Reteplase (international nonproprietary name)
Description:	Rapilysin (Retavase in the U.S.) is a recombinant nonglycosylated engineered human tissue plasminogen activator (tPA). It contains two of the five domains of the natural protein and retains the ability to cleave plasminogen into plasmin. Rapilysin is produced in *Escherichia coli* by recombinant DNA technology and is presented in a lyophilized form to be reconstituted before intravenous administration using the provided solvent-filled syringe.
Approval Date:	1996 (E.U. and U.S.)
Therapeutic Indications:	Rapilysin is indicated in the thrombolytic therapy of acute myocardial infarction.
Manufacturer:	Roche Diagnostics GmbH, Sandhofer Strasse 116, 68305 Mannheim, Germany, http://www.roche-diagnostics.com (E.U. and U.S.)
Marketing:	Roche Registration Limited, 40 Broadwater Road, Welwyn Garden City, Hertfordshire AL7 3AY, U.K., http://www.roche.com (E.U.) Centocor, Inc., 200 Great Valley Parkway, Malvern, PA 19355, http://www.centocor.com (U.S.)

Manufacturing

Rapilysin is a recombinant nonglycosylated form of the human tPA. It consists of the kringle-2 domain, related to the activity of the molecule, and the serine protease domain of the natural protein, while lacking the kringle-1, finger, and epidermal growth factor domains of native tPA. The modified gene is encoded on a plasmid that is transformed into *E. coli* cells, with a helper vector used to increase the stability of the construct. The protein is expressed in the bacterial strain and aggregates in the cells in the form of inclusion bodies. After isolation of inclusion bodies, protein solubilization and denaturation are carried out under reducing conditions, followed by renaturation in the presence of glutathione in order to obtain the active protein. Misfolded molecules are removed via chromatographic steps and filtration procedures. The final formulation includes the active substance reteplase and the following excipients: tranexamic acid di-potassium-hydrogen phosphate, phosphoric acid, sucrose, and polysorbate 80. The product is presented in a lyophilized form to be reconstituted using the solvent provided in a sterile, prefilled syringe before intravenous administration. The shelf life of the product is 24 months when stored below 25°C.

Overview of Therapeutic Properties

Rapilysin, a thrombolytic agent, is indicated for the treatment of acute myocardial infarction. Thrombolytics dissolve clots in blood vessels, which are the cause of heart attacks. Rapilysin catalyses the degradation of plasminogen to plasmin, which degrades the fibrin matrix of the clot. The deletion of three of the domains of the natural plasminogen activator, from which it is derived, and the lack of glycosylation give Rapilysin a faster plasma clearance and a shorter half-life. This allows for a two-dose administration with a 30-minute interval (which is significantly shorter than the 90-minute interval observed with the natural molecule) and for increased efficacy. Furthermore, Rapilysin binds to fibrin less tightly than the natural molecule, thus allowing better diffusion throughout the clot.

Rapilysin should be administered within 12 hours of the occurrence of heart attack symptoms, with heparin and acetylsalicylic acid also given both before and after Rapilysin. Because Rapilysin interacts with heparin, care should be taken to keep administrations separate.

Rapilysin proved to be at least as effective and safe as the other available thrombolytic agents, altepase and streptokinase. No studies have been carried out on patients younger than 18 years old. Rapilysin should not be administered in the case of bleedings or during pregnancy and lactation.

Further Reading

http://www.centocor.com
http://www.eudra.org
http://www.fda.gov
http://www.retavase.com

Askari, A.T. and Lincoff, A.M., GUSTO V: combination drug treatment of acute myocardial infarction. Global use of strategies to open occluded coronary arteries, *Cleve. Clin. J. Med.*, 69, 554–560, 2002.

Noble, S. and McTavish, D., Reteplase. A review of its pharmacological properties and clinical efficacy in the management of acute myocardial infarction. *Drugs,* 52, 589–605, 1996.

Waller, M. and Kohnert, U., Reteplase, a recombinant plasminogen activator, in *Biopharmaceuticals, an Industrial Perspective,* Walsh, G. and Murphy, B., Eds., Dordrecht: Kluwer Academic Publisher, 185–216, 1999.

Wooster, M.B. and Luzier, A.B., Reteplase: a new thrombolytic for the treatment of acute myocardial infarction, *Ann. Pharmacother.*, 33, 318–324, 1999.

Rebetron

Product Name: Rebetron (combination therapy) (trade name)

Ribavirin capsules in combination with interferon alfa-2b, recombinant, injection (common name)

Description: Rebetron is a combination therapy in which target patients are administered a combination of Intron A (Schering Corporation's brand name for recombinant interferon α-2b; see Intron A monograph) and Rebetol (Schering Corporation's brand name for ribavirin). Ribavirin (1-β-D-ribofuranosyl-1H-1,2,4-triazole-3 carboxamide; $C_8H_{12}N_4O_5$) is a nucleoside analogue with antiviral activity. The product is supplied as a combination package containing rebetol capsules and intron A injection provided in either single-use or multidose vials or in a multidose pen.

Approval Date: 1999 (U.S.)

Therapeutic Indications: Treatment of chronic hepatitis C in patients with compensated liver disease previously untreated with interferon alfa-2b or who have relapsed following interferon alfa-2b therapy.

Manufacturer: Schering-Plough Corporation, 2000 Galloping Hill Road, Kenilworth, NJ 07033, http://www.schering.com

Marketing: Schering-Plough Corporation, 2000 Galloping Hill Road, Kenilworth, NJ 07033, http://www.schering.com

Manufacturing

See Intron A monograph.

Overview of Therapeutic Properties

Treatment of the target population with rebetron resulted in positive virological and histological (liver biopsy) responses in some patients. The mechanism of inhibition of the hepatitis C virus by the combination therapy has not been established. The therapeutic regimen for adults normally entails daily consumption of two 200-mg Rebetol capsules in the morning and three in the evening, along with subcutaneous injections of 3 million IU of Intron A three times a week.

Significant teratogenic and embryocidal effects have been demonstrated with ribavirin in all animal species studied. Rebetron combination therapy is contraindicated in females who are pregnant and in male partners of pregnant females. Extreme care must be taken to avoid pregnancy during treatment and for 6 months following treatment. Additional adverse effects include hemolytic anemia, bone marrow toxicity, and psychiatric adverse events.

Further Reading

1. Herrine, S.K. et al., Efficacy and safety of Peginter from alfa 2a pegasys combination therapies in patients who relapsed on Rebetron therapy, *Hepatology,* 36(4, Part 2 Suppl.), 781, 2002.
2. Liu, J.J. et al., Succesful treatment of giant cell hepatitis with Rebetron (interferon/ribavirin), *Am. J. Gastroenterol.,* 98(1), 223–224, 2003.

Rebif

Product Name:	Rebif (trade name) Interferon beta-1a (international nonproprietary name)
Description:	Rebif is a recombinant human interferon beta-1a (IFN-β-1a) produced by recombinant DNA technology in Chinese hamster ovary (CHO) cells. It is presented as a solution in a prefilled syringe (2 strengths; 22 or 44 μg — equivalent to 6 or 12 million IU activity/0.5 ml) for subcutaneous administration.
Approval Date:	1998 (E.U.); 2002 (U.S.)
Therapeutic Indications:	Rebif is indicated for treatment of patients with relapsing-remitting multiple sclerosis.
Manufacturer:	Serono Pharma S.p.A., Via de Blasio, Zona Industriale di Modugno, 70123 Bari, Italy, http://www. serono. com (manufacturer is responsible for batch release in the European Economic Area) Serono, Inc., One Technology Place, Rockland, MA 02370, http://www.seronousa.com (U.S.)
Marketing:	Serono Europe Limited, 56 Marsh Wall, London E14 9TP, U.K., http://www.serono.com (E.U.) Serono, Inc., One Technology Place, Rockland, MA 02370, http://www.seronousa.com (U.S.)

Manufacturing

Rebif is a recombinant human IFN-β-1a. The gene has been cloned into a modified CHO cell line, and the product, similar to the natural interferon, is purified from the culture medium. Purification involves a number of chomatographic steps (affinity, ion-exchange, size exclusion, and reversed-phase liquid chromatography) and filtration procedures. The final product consists of IFN-β-1a (active), as well as mannitol, human serum albumin, sodium acetate, acetic acid, and sodium hydroxide as excipients. Rebif is provided as a solution in a prefilled syringe for subcutaneous administration.

The shelf life of the product is 24 months when stored at 2 to 8°C. Extensive tests performed during processing and on the final product include SDS-PAGE, RT-HPLC, SE-HPLC, IE HPLC, N-terminal amino acid sequencing, peptide mapping, carbohydrate mapping, and assays to ensure viral and microbial safety.

Overview of Therapeutic Properties

Multiple sclerosis is a chronic disease affecting the central nervous system. The only treatment currently available merely alleviates the symptoms of the disease and reduces the frequency of relapses and disease progression. Interferons mediate signaling among cells and are involved in the immunoresponse. Treatment with interferons reduces the progression of multiple sclerosis, but the mechanism by which this occurs is unclear. Rebif is indicated in the treatment of patients with relapsing-remitting multiple sclerosis who have had at least two relapsing episodes in the previous 2 years. Rebif should be administered subcutaneously three times a week.

The efficacy and safety of Rebif has been established in clinical trials. Patients receiving Rebif exhibited reduced frequency and severity of relapses more than did patients receiving placebos, leading to a slowing down of progression of the disease. The most common side effects reported after administration were flu-like symptoms and necrosis at the sites of injection. Antibodies against IFN-β-1a developed in 25% of patients after 1 to 2 years of treatment, leading in some cases to a reduced efficacy of the treatment. Rebif should not be administered in case of depression, epileptic disorder, and during pregnancy and lactation. No studies have been carried out on patients younger than 16 years old.

Further Reading

http://www.eudra.org
http://www.fda.gov
http://www.rebif.com
http://www.serono.com
Bagnato, F. et al., A one-year study on the pharmacodynamic profile of interferon beta-1a in MS, *Neurology*, 58, 1409–1411, 2002.
Beck, R.W. et al., Interferon beta-1a for early multiple sclerosis: CHAMPS trial subgroup analyses, *Ann. Neurol.*, 51, 481–490, 2002.

Recombinate

Product Name:	Recombinate (trade name) Antihaemophilic factor (recombinant) (common name)
Description:	Recombinate consists of a recombinant human coagulation factor VIII. It is produced in Chinese hamster ovary (CHO) cells using recombinant DNA technology. It is provided as a lyophilized powder to be reconstituted before intravenous administration.
Approval Date:	1992 (U.S.)
Therapeutic Indications:	Recombinate is indicated for the prevention and control of bleedings in patients suffering from hemophilia A. It is also indicated in the prophylaxis for surgery.
Manufacturer:	Baxter Healthcare Corporation, Hyland Immuno, Glendale, CA 91203, http://www.baxter.com
Marketing:	Baxter Healthcare Corporation, Hyland Immuno, Glendale, CA 91203, http://www.baxter.com

Manufacturing

The recombinant antihemophilic factor (rAHF) is a glycoprotein very similar to the naturally occurring human coagulation factor VIII. It is produced in CHO cells using recombinant DNA technology. The glycoprotein is co-expressed with von Willebrand's factor, which stabilizes rAHF in a noncovalently linked complex during production. Purification involves a number of chromatographic steps, including immunoaffinity chromatography using murine monoclonal antibodies against the heavy chain of rAHF. The purified rAHF consists of a combination of heterologous heavy and light chains very similar to the natural human glycoprotein. The final product consists of rAHF, human serum albumin (as a stabilizer), calcium, polyethylene glycol, sodium, histidine, polysorbate 80, traces of bovine and hamster proteins from the culture, murine protein from the purification, and von Willebrand's factor.

Recombinate is presented as a lyophilized powder to be reconstituted with the supplied solvent before intravenous administration. Extensive tests are carried out to validate the potency, identity, purity, sterility, viral safety, and specific activity of the final product.

Overview of Therapeutic Properties

The antihemophilic factor, or coagulation factor VIII, is involved in the coagulation process, a cascade of reactions resulting in the formation of a fibrin clot. Patients suffering from hemophilia A lack a functional factor VIII and therefore require replacement therapy in order to avoid and control bleedings. The use of a recombinant form of the coagulation factor offers the advantages of unlimited availability and the elimination of risks related to blood-derived products over the plasma-derived protein.

Recombinate is indicated for hemophilia A in patients of all ages, and it should be administered intravenously at a dosage adjusted to individual needs. Recombinate was shown to be as effective as the natural human coagulation factor VIII in the management of hemophilia A.

Recombinate was found to be safe, with a very low incidence of side effects, which were mostly infusion-related. Allergic reactions, due to traces of murine, bovine, and hamster proteins, were rarely reported. Neutralizing antibodies that interfere with the efficacy of the product have been reported. Recombinate should be administered during pregnancy only if clearly necessary.

Further Reading

http://www.baxter.com

http://www.fda.gov

Bray, G.L. et al., A multicenter study of recombinant factor VIII (recombinate): safety, efficacy, and inhibitor risk in previously untreated patients with hemophilia A, *Blood*, 83, 2428–2435, 1994.

Ingerslev, J. et al., Antibodies to heterologous proteins in hemophilia A patients receiving recombinant factor VIII (Recombinate), *Thromb. Haemost.*, 87, 626–634, 2002.

Scoble, H.A., Recombinate antihaemophilic factor, *Dev. Biol. Stand.*, 96, 141–147, 1998.

White, G.C. et al., A multicenter study of recombinant factor VIII (recombinate) in previously treated patients with hemophilia A, *Thromb. Haemost.*, 77, 660–667, 1997.

Recombivax

Product Name:	Recombivax (trade name) Hepatitis B (recombinant) vaccine (common name)
Description:	Recombivax is a vaccine against the hepatitis B virus. It contains the major surface antigen of the hepatitis B virus produced by recombinant DNA technology in *Saccharomyces cerevisiae*. Recombivax is presented as a suspension for intramuscular injection in vials or prefilled syringes.
Approval Date:	1986. Preservative-free formulation for infants approved in 1999. Two-dose regimen for adolescents approved in 2000 (U.S.).
Therapeutic Indications:	Recombivax is indicated for immunization against hepatitis B.
Manufacturer:	Merck and Co., Inc., P.O. Box 4, Sumneytown Pike, West Point, PA 19486, http://www.merck.com
Marketing:	Merck and Co., Inc., P.O. Box 4, Sumneytown Pike, West Point, PA 19486, http://www. merck.com/

Manufacturing

Recombivax is a recombinant vaccine that contains the major surface antigen of the hepatitis B virus. The protein is produced in a modified *S. cerevisiae* strain using recombinant DNA technology. It differs from the natural viral molecule in lacking glycosylation, but it retains its immunogeniciy. The protein is purified using several chromatographic and filtration procedures, following which the purified protein is inactivated using formaldehyde and adsorbed on aluminum hydroxide, which acts as an adjuvant. The final product is presented as a suspension for intramuscular administration and is available in separate adult, pediatric and adolescent, and dialysis formulations. All the formulations contain the major hepatitis B surface antigen adsorbed to aluminium hydroxide, and they are available with or without a preservative (thiomersal). The quality and safety of the vaccine are ensured by extensive testing for sterility, stability, identity, potency, and the presence of contaminants.

Overview of Therapeutic Properties

Recombivax is a recombinant vaccine indicated for the immunization against the hepatitis B virus. Different formulations are available for infants, children, adolescents (11 to 15 years of age), adults, and predialysis and dialysis adults. Recombivax should be administered intramuscularly in the deltoid muscle, or in the anterolateral thigh in the case of infants. A three-dose regimen is recommended, with the second dose administered 1 month and the third dose administered 6 months after the first. A two-dose regimen in adolescents, with the second injection 4 to 6 months after the first, was found to induce the same immunoprotection as the standard three-dose regimen. Predialysis and dialysis patients may require a booster dose for full immunoprotection.

Recombivax was found to efficiently elicit immune protection against the hepatitis B virus. It was also found that immunization with Recombivax reduced the incidence of liver cancer. Recombivax was demonstrated to be safe and well tolerated, with reactions at the sites of injection, flu-like symptoms, and irritability the most commonly reported side effects.

Recombivax may be administered with diphtheria, tetanus, pertussis, polio, measles, mumps, rubella, and hemophilus b vaccines. Care should be taken when Recombivax is administered during pregnancy and lactation, and it should not be administered in cases of fever.

Further Reading

http://www.fda.gov

http://www.merck.com

Cassidy, W.M. et al., A randomized trial of alternative two- and three-dose hepatitis B vaccination regimens in adolescents: antibody responses, safety, and immunologic memory, *Pediatrics*, 107, 626–631, 2001.

Dukes, C.S. et al., Hepatitis B vaccination and booster in predialysis patients: a 4-year analysis, *Vaccine*, 11, 1229–1232, 1993.

ReFacto

Product Name: ReFacto (trade name)

Moroctocog alfa (international nonproprietary name)

Description: ReFacto is a recombinant coagulation Factor VIII. It is a glycosylated protein that lacks most of the B domain of the natural protein but retains the same anticoagulation activity. Structurally, it comprises a total of 1438 amino acid residues, which are composed of two glycosylated polypeptide chains (80 and 90 kDa, respectively) held together by a metal ion bridge. It is produced in Chinese hamster ovary (CHO) cells by recombinant DNA technology, and it is provided in a lyophilized form to be reconstituted before intravenous administration.

Approval Date: 1999 (E.U.); 2000 (U.S.)

Therapeutic Indications: ReFacto is indicated for the treatment and prophylaxis of bleeding in patients with hemophilia A.

Manufacturer: Wyeth Laboratories, New Lane, Havant, Hants, PO9 2NG, U.K., http://www.wyeth.com (manufacturer is responsible for batch release in the European Economic Area)

Wyeth Pharmaceuticals, Inc., One Burtt Road, Andover, MA 01810-5901 (U.S.), http://www.wyeth.com

Marketing: Genetics Institute of Europe B.V., Fraunhoferstrasse 15, 82152 Planegg/Martinsried, Germany, http://www.genetics.com (E.U.)

Wyeth Pharmaceuticals, Inc., One Burtt Road, Andover, MA 01810-5901 (U.S.), http://www.wyeth.com

Manufacturing

ReFacto consists of a recombinant coagulation factor VIII that lacks most of the B domain of the natural protein, thus reducing the size of the natural molecule by about 40%. The recombinant protein retains the activity of the natural protein but has the advantage of being easier to produce. The gene encoding the recombinant molecule is cloned, and the protein is expressed in CHO cells. After processing and glycosylation, the protein is secreted into the culture medium. Purification involves chromatographic steps, including immunoaffinity chromatography using a murine monoclonal antibody raised against the protein. Viral inactivation steps are also included during downstream processing. The purified product is lyophilized. No human serum albumin is added for stabilization, unlike other formulations of factor VIII, due to the increased stability of the recombinant protein. The final formulation contains moroctocog alfa, the active substance, and the following excipients: sucrose, calcium chloride, L-histidine, polysorbate 80, and sodium chloride. The lyophilized product is administrered intravenously after reconstitution using the provided solvent.

The shelf life of the product is 24 months when stored at 2 to 8°C. Validation tests to ensure the quality and safety of the product include SDS-PAGE for identity, SEC-HPLC for the presence of aggregates and fragments, and tests for sterility and the presence of contaminants and endotoxins.

Overview of Therapeutic Properties

The coagulation factor VIII is involved in the blood coagulation process. Once activated, it accelerates the conversion of factor X by acting as a cofactor for factor IX. Factor X activates prothrombin to form thrombin, which is necessary to convert fibrinogen into fibrin to form the clot.

Patients with hemophilia A do not have a functional factor VIII, due to either a lack of the protein or insufficient activity. A replacement therapy is therefore necessary to avoid prolonged bleeding episodes. The use of the recombinant product offers the advantage over the plasma-derived factor VIII of increased availability and of eliminating the risks of accidental disease transmission. The increased stability of the recombinant protein allows omission of human serum albumin (HSA) as a stabilizer. HSA is generally added to formulations of factor VIII.

ReFacto may be administered to patients of all age groups with the dosage adjusted to meet individual needs. ReFacto has proven effective in either previously treated or untreated hemophilia A patients and may be used in prophylaxis for surgery. The most common side effects are those developed after intravenous administration of proteins. Some allergic reactions have been observed, and neutralizing antibodies have been reported in 20 to 30% of patients, which was also observed after the use of other factor VIII products.

Further Reading

http://www.eudra.org
http://www.fda.gov
http://www.genetics.com
http://www.wyeth.com
Courter, S.G. and Bedrosian, C.L., Clinical evaluation of B-domain deleted recombinant factor VIII in previously treated patients, *Semin. Hematol.*, 38(2 Supp. 4), 44–59, 2001.
Sandberg, H. et al., Structural and functional characteristics of the B-domain-deleted recombinant factor VIII protein, r-VIII SQ, *Thromb. Haemost.*, 85, 93–100, 2001.

Refludan

Product Name:	Refludan (trade name) Lepirudin (international nonproprietary name)
Description:	Refludan is a recombinant hirudin analogue that acts as an inhibitor of thrombin. It is produced by recombinant DNA technology in *Saccharomyces cerevisiae* cells. The 65 amino acid peptide differs from the natural leech hirudin molecule at the N-terminal end, where an isoleucine residue replaces a leucine, and in the absence of a sulfate group on the tyrosine at position 63. Refludan is supplied in a lyophilized form (50 mg/vial) to be reconstituted before intravenous administration.
Approval Date:	1997 (E.U.); 1998 (U.S.)
Therapeutic Indications:	Refludan is indicated for anticoagulation in adult patients with heparin-associated thrombocytopenia (HAT) and associated thromboembolic disease.
Manufacturer:	Hoechst Marion Roussel Deutschland GmbH, Marburg, Lahn, Germany (manufacturer is responsible for import and batch release in the European Economic Area), http:www.aventis.com Aventis Behring Deutschland GmbH, D-35002, Marburg, Germany (U.S.), http:www.aventisbehring.de
Marketing:	Schering AG, 13342 Berlin, Germany, http://www.schering.de (E.U.) Berlex Laboratories, 6 West Belt Road, Wayne, NJ 07470 (U.S.), http:www.berlex.com

Manufacturing

Lepirudin is a hirudin analogue that is produced by recombinant DNA technology. The synthetic gene is derived from the sequence of the hirudin polypeptide found in salivary secretions from the leech *Hirudo medicinalis.*

Lepirudin is produced in a modified *S. cerevisiae* strain. The purification process includes a number of chromatographic steps and ultrafiltration-diafiltration.

The final product is supplied in a lyophilized form consisting of lepirudin (active), as well as mannitol (a bulking and tonicity agent) and sodium hydroxide. The shelf life of the product is 24 months when stored below 25°C. The product undergoes extensive control tests to ensure quality and safety, such as PR-HPLC, amino acid composition and peptide mapping, assays for clarity of solution, and tests for the presence of yeast and microbial contamination proteins and bacterial endotoxins.

Overview of Therapeutic Properties

Refludan is an analogue of hirudin that is produced naturally in leech salivatory gland cells. It functions to prevent blood coagulation during the consumption of human blood. Hirudin has been used for centuries for its anticoagulation properties. The use of recombinant DNA technology enables the production of great quantities of lepirudin for therapeutic use. Lepirudin acts as a direct inhibitor of thrombin and blocks the formation of the fibrin clot. Patients treated with heparin over a prolonged period can develop HAT, characterized by a reduction of platelets in the blood and an increased risk of clotting. No treatment is available for HAT other than discontinuation of administration of heparin. Lepirudin, with a different mechanism of action from heparin, can be used in those cases in order to avoid formation of blood clots and allow the recovery of blood platelets.

Refludan proved to be effective as an anticoagulant in patients affected by HAT, leading to increased survival. The major side effect after treatment with Refludan appeared to be bleeding, which affected 11% of patients. Rare allergic episodes have also been reported. The formation of antilepirudin antibodies was observed in 40% of patients. These antibodies showed no neutralizing effect; on the contrary, they led to an increased efficacy of the product probably due to slow elimination of lepirudin when complexed with the antibody. Refludan is contraindicated during pregnancy and lactation.

Further Reading

http://www.aventis.com
http://www.eudra.org
http://www.fda.gov
http://www.refludan.com

Greinacher, A. et al., Lepirudin (recombinant hirudin) for parenteral anticoagulation in patients with heparin-induced thrombocytopenia, *Circulation*, 100, 587–593, 1999.

Greinacher, A. et al., Heparin-induced thrombocytopenia with thromboembolitic complications: meta-analysis of two prospective trials to assess the value of parenteral treatment with lepirudin and its therapeutic aPTT range, *Blood*, 96, 846–851, 2000.

Warkentin, T.E. and Kelton, J.G., A 14-year study of heparin-induced thrombocytopenia, *Am. J. Med.*, 101, 502-507, 1996.

Regranex

Product Name:	Regranex (trade name) Becaplermin (international nonproprietary name)
Description:	Regranex is a recombinant human platelet-derived growth factor (PDGF). It is a homodimer. Each 109-amino acid, glycosylated polypeptide is aligned in an antiparallel fashion relative to the other. They are held in position by two interchain disulphide bonds, yielding the 24.5-kDa mature molecule. Regranex is produced by recombinant DNA technology in the yeast *Saccharomyces cerevisiae*. The product is presented in a gel formulation containing 0.1% active ingredient for external topical use.
Approval Date:	1997 (U.S.); 1999 (E.U.)
Therapeutic Indications:	Regranex is indicated for the treatment of neuropathic diabetic ulcers.
Manufacturer:	Janssen Pharmaceutica N.V., Turnhoutseweg 30, 2340 Beerse, Belgium, http://www.janssenpharmaceutica.be (manufacturer is responsible for batch release in the European Economic Area) OMJ Pharmaceuticals, Inc., Road 362, Km 0.5, San German, Puerto Rico 00683 (U.S.)
Marketing:	Janssen-Cilag International N.V., Turnhoutseweg 30, 2340 Beerse, Belgium, http://www.janssen-cilag.com (E.U.) Ortho -McNeil Pharmaceuticals, Inc., 1000 U.S. Route 202, Raritan, NJ 08869, http://www.ortho-mcneil.com (U.S.)

Manufacturing

The recombinant human PDGF becaplermin is produced by recombinant DNA technology in a modified *S. cerevisiae* strain containing the gene encoding the B chain. The glycosylated protein is extracted from the culture and undergoes purification steps, including multiple chromatographic and ultrafiltration procedures. The final product is presented as a gel containing the active substance, becaplermin, and the following excipients: carmellose sodium, sodium chloride, sodium acetate, glacial acetic acid, methyl parahydroxybenzoate (methylparaben, as a preservative), propyl parahydroxybenzoate (propylparaben, as a preservative), m-cresol (as a preservative), and lysine hydrochloride. The product is a nonsterile gel and is used only for external cutaneous application.

The shelf life of the product is 12 months when stored at 2 to 8°C. Validation tests to ensure the quality and safety of the product are carried out during production and on the final product.

Overview of Therapeutic Properties

Diabetic patients frequently develop skin ulcers that are generally treated using normal wound care practices. In some cases this fails to heal the wound, which can lead to infections and, eventually, amputation. The active substance of Regranex, the recombinant human PDGF, accelerates the healing process by inducing cell growth and tissue repair. Regranex is indicated for the treatment of chronic diabetic ulcers that do not heal with normal wound care practice. Regranex should be administered daily and up to a maximum of 20 weeks. The wound, if infected, should be cleaned before application of the gel, and no weight should be applied to it.

Regranex proved to be efficient in healing chronic diabetic ulcers of less than 5 cm^2. Rarely, skin irritations were reported after application, probably due to the preservatives in the preparation. No studies have been carried out on patients younger than 18 years old. Regranex should not be administered during pregnancy and lactation or in the case of malignancies.

Further Reading

http://www.eudra.org
http://www.fda.gov
http://www.jnj.com
http://www.ortho-mcneil.com
http://www.regranex.com

Embil, J.M. and Nagai, M.K., Becaplermin: recombinant platelet derived growth factor, a new treatment for healing diabetic foot ulcers, *Expert Opin. Biol. Ther.*, 2, 211–218, 2002.

Piascik, P., Use of Regranex gel for diabetic foot ulcers, *J. Am. Pharm. Assoc.*, 38, 628–630, 1998.

Remicade

Product Name: Remicade (trade name)

Infliximab (international nonproprietary name)

Description: Remicade is a chimeric (murine-human) monoclonal antibody with specific binding activity for tumor necrosis factor-α (TNF-α), a cytokine involved in inflammatory and immune responses. It is produced by recombinant DNA technology and supplied in a lyophilized form (100 mg active/vial) to be resuspended as a concentrated solution for intravenous infusion.

Approval Date: 1998 for treatment of Crohn's disease, 1999 for reduction of signs and symptoms of rheumatoid arthritis, 2000 for inhibition of progression of structural damage in rheumatoid arthritis (U.S.); 1999 for treatment of Crohn's disease, 2000 for treatment of rheumatoid arthritis (E.U.)

Therapeutic Indications: Remicade is indicated in the treatment of severe, active Crohn's disease and fistulizing Crohn's disease in patients who have not responded to conventional therapies. It is indicated for the reduction of signs and symptoms of rheumatoid arthritis as well as the inhibition of progression of structural damage in patients who have not responded to other drugs, including methotrexate.

Manufacturer: Centocor B.V., Einsteinweg 101, 2333 CB Leiden, the Netherlands, http://www.centocor.com (E.U.)
Centocor, Inc., 200 Great Valley Parkway, Malvern, PA 19355 (U.S.), http://www.centocor.com

Marketing: Centocor B.V., Einsteinweg 101, 2333 CB Leiden, the Netherlands, http://www.centocor.com (E.U.)
Centocor, Inc., 200 Great Valley Parkway, Malvern, PA 19355 (U.S.), http://www.centocor.com

Manufacturing

The chimeric monoclonal antibody Remicade (30% murine, 70% human) consists of murine variable regions and heavy and kappa light constant regions from a human IgG1. The murine hybridoma was obtained after immunization of mice with the human TNF-α. The recombinant antibody is produced in and secreted from murine SP0/2 cells. The purification process involves various chromatographic steps, including affinity and anion exchange chromatography and viral inactivation and removal procedures. The final product is supplied as a lyophilized form of infliximab and includes sucrose, polysorbate 80, monobasic sodium phosphate, and dibasic sodium phosphate as excipients.

The shelf life of the product is 24 months when stored at 2 to 8°C. Routine evaluative tests on the final product include isoelectric focusing, GF-HPLC, SDS-PAGE, tests for the presence of entotoxins, bioburden, and for viral and bacterial safety.

Overview of Therapeutic Properties

Remicade is indicated as a treatment to interrupt the process of inflammation. It binds specifically and with high affinity to TNF-α, a cytokine that mediates the inflammation process and the immune response. Remicade reduces the accumulation of TNF-α in the joints of patients affected by rheumatoid arthritis, reducing inflammation. The same anti-inflammatory effect is observed in Crohn's disease, a chronic incurable inflammatory bowel disease that could develop complications such as the occurrence of fistula. Remicade is used as a treatment for these diseases when conventional therapies, including corticosteroids and immunosuppressive drugs, have had no adequate response.

Remicade is administered as an infusion over a 2-hour period. Additional doses are administered 2 and 6 weeks after the first dose. In rheumatoid arthritis patients, a dose is then administered after every 8 weeks; methotrexate is given in combination with Remicade as part of the treatment. In Crohn's disease patients the treatment may be repeated, if necessary, but within 14 weeks of the previous dose. Administration after a longer drug-free period of time may cause severe hypersensitivity reactions.

Clinical trials in Crohn's disease patients showed that Remicade reduced the activity of the disease in 60% of the patients and also proved to be effective in fistulising disease. A reduction of signs and symptoms of rheumatoid disease was shown in clinical trials in which patients were treated with Remicade in combination with methotrexate.

Allergic reactions, generally mild, may occur during infusion. The most common side effects associated with the treatment were upper respiratory infections, headaches, nausea, sinusitis, rashes, and cough. The blockade of TNF-α increases the risk of infections, and tuberculosis and other opportunistic infections have been reported, some of them fatal. Immunogenic reactions with the production of human antichimeric antibodies have been observed. Antinuclear antibodies and antidouble-stranded DNA antibodies may appear during the treatment, but only rarely develop an autoimmune lupus-like syndrome. Remicade should not be administered to patients with congestive heart failure and is contraindicated during pregnancy.

Further Reading

http://www.centocor.com
http://www.eudra.org
http://www.fda.gov
http://www.remicade.com
Elliott, M.J. et al., Treatment of rheumatoid arthritis with chimeric monoclonal antibodies to tumor necrosis factor α, *Arthritis Rheum.*, 36, 1681–1690, 1993.
Valle, E. et al., Infliximab, *Expert. Opin. Pharmacother.*, 2, 1015–1025, 2001.
van Dullemen, H.M. et al., Treatment of Crohn's disease with antitumor necrosis factor chimeric monoclonal antibody (cA2), *Gastroenterology*, 109,129–135, 1995.

ReoPro

Product Name: ReoPro (trade name)

Abciximab (international nonproprietary name)

Description: Abciximab is a recombinant chimeric human-murine Fab antibody fragment. It consists of the variable fragments of the murine monoclonal antibody 7E3 combined with constant regions from human-derived light and heavy chains. The Fab fragment is specific for the glycoprotein IIb/IIIa receptor, which is found on human platelets, and binding inhibits platelet aggregation. Abciximab is produced by continuous perfusion culture in a modified murine myeloma Sp2/0 cell line. It is provided as a solution (2 mg/ml) for intravenous injection and infusion.

Approval Date: 1994 (U.S.)

Therapeutic Indications: Abciximab is indicated as an adjunct to percutaneous coronary intervention for the prevention of cardiac ischemic complications in patients undergoing percutaneous coronary intervention. It is also indicated in patients with unstable angina who have not responded to conventional medical therapy and in whom percutaneous coronary intervention is planned within 24 hours. Abciximab should be used in combination with aspirin and heparin.

Manufacturer: Centocor B.V., Einsteinweg 101, P.O. Box 251, 2300 AG Leiden, the Netherlands, http://www.centocor.com

Marketing: Centocor B.V., Einsteinweg 101, P.O. Box 251, 2300 AG Leiden, the Netherlands, http://www.centocor.com, and distributed by Eli Lilly and Company, Lilly Corporate Center, Indianapolis, IN 46285, http://www.lilly.com

Manufacturing

Abciximab is a chimeric human-murine Fab antibody fragment. The heavy and light variable regions are derived from the murine monoclonal antibody 7E3, which is specific for the glycoprotein IIb/IIIa receptor found on human platelets. 7E3 was obtained from mice immunized with human platelets, following which the variable regions were combined with constant regions of human origin. The chimeric human-murine antibody is produced by continuous perfusion culture in a modified murine myeloma S2/0 cell line. The antibody is purified, and the Fab fragment of the antibody is obtained by digestion with papain. Purification involves several chromatographic steps and viral removal procedures. The final product is provided as a liquid formulation that contains the antibody fragment abciximab, as well as sodium phosphate, sodium chloride, and polysorbate 80 as excipients.

Overview of Therapeutic Properties

Percutaneous coronary intervention is performed to enlarge the lumen of stenosed arteries. As a consequence of the intervention, there is a high risk of thrombus formation, leading to coronary occlusion. ReoPro is used in combination with aspirin and heparin to reduce the risk of thrombosis. The chimeric antibody fragment binds to the glycoprotein IIb/IIIa receptor on the surface of platelets and inhibits the binding of extracellular matrix proteins (e.g., fibrinogen and von Willebrand factor), thus inhibiting platelet aggregation and clot formation. The antibody also binds to the vitronectin receptor.

ReoPro is administered as an intravenous bolus 10 to 60 minutes before the percutaneous coronary intervention, followed by an intravenous infusion for 12 hours. Clinical trials were performed on a broad population of patients undergoing percutaneous coronary interventions, including balloon angioplasty, atherectomy, and stent implantation, as well as patients with unstable angina who did not respond to conventional medical treatment. Heparin and aspirin were used in combination with ReoPro. Studies showed an improvement of up to 50% in patients treated with ReoPro compared to patients treated with placebos in terms of death, myocardial infarctions, and urgent interventions at 30 days, 60 days, and up to 3 years after the percutaneous coronary intervention.

The most frequent side effect reported after the use of ReoPro was bleeding, most commonly at the site of intervention, which could be minimized to the same extent as in patients receiving placebos by adjusting the level of heparin administered to suit individual body weights. ReoPro should be used with extreme care during pregnancy and lactation. No studies have been performed on the pediatric use of ReoPro. Ongoing studies indicate the potential for use of ReoPro in combination with fibrinolytic therapy.

Further Reading

http://www.abciximab.com
http://www.centocor.com
http://www.fda.gov
http://www.reopro.com
Cheng, J.W., Efficacy of glycoprotein IIb/IIIa-receptor inhibitors during percutaneous coronary intervention, *Am. J. Health Syst. Pharm.*, 59(21 Suppl. 7), 5–14, 2002.
Jordan, R.E., Nakada, M.T., and Weisman, H.F., Abciximab: the first platelet glycoprotein IIb/IIIa receptor antagonist, in *Biopharmaceuticals, an Industrial Perspective,* Walsh, G. and Murphy, B., Eds., Dordrecht, the Netherlands: Kluwer Academic Publisher, 35–71, 1999.
Kandzari, D.E. and Califf, R.M., TARGET versus GUSTO-IV: appropriate use of glycoprotein IIb/IIIa inhibitors in acute coronary syndromes and percutaneous coronary intervention, *Curr. Opin. Cardiol.*, 17, 332–339, 2002.
Ottervanger, J.P. et al., Long-term results after the glycoprotein IIb/IIIa inhibitor abciximab in unstable angina: one-year survival in the GUSTO IV-ACS (Global Use of Strategies To Open Occluded Coronary Arteries IV — Acute Coronary Syndrome) Trial, *Circulation,* 107, 437–442, 2003.

Replagal

Product Name:	Replagal (trade name) Agalsidase alfa (international nonproprietary name)
Description:	Replagal is a recombinant human α-galactosidase A, a lysosomal hydrolase. The enzyme is a glycosylated homodimer, with each 50-kDa, 398-amino acid subunit containing three N-linked oligosaccharides. It is produced in a continuous human cell line by recombinant DNA technology, and it is supplied as a concentrated solution (1 mg/ml) for intravenous infusion over 40 minutes.
Approval Date:	2001 (E.U.)
Therapeutic Indications:	Long-term enzyme replacement therapy in patients with Fabry's disease (α-galactosidase A deficiency).
Manufacturer:	TKT Europe-5S AB, Rinkebyvägen 11B, SE 182 36 Danderyd, Sweden, http://www.tktx.com (manufacturer is responsible for import and batch release in the European Economic Area)
Marketing:	TKT Europe-5S AB, Rinkebyvägen 11B, SE 182 36 Danderyd, Sweden, http://www.tktx.com

Manufacturing

Replagal is a recombinant human α-galactosidase A, a lysosomal hydrolase. It is produced by recombinant DNA technology in a continuous human cell line. The continuous human cell line was chosen in order to obtain the same glycosylation pattern as displayed by the natural molecule. The recombinant protein is purified using a combination of five chromatographic steps and a viral removal procedure (filtration). The final product contains agalsidase alfa, the active substance, and the following excipients: monohydrate monobasic sodium phosphate, polysorbate 20, sodium chloride, and sodium hydroxide. It is offered as a concentrated solution to be diluted before administration as an intravenous infusion.

The shelf life of the product is 12 months when stored at 2 to 5°C. Tests carried out to characterize the recombinant molecule and to ensure the quality and safety of the product include isoelectric focusing, Western blot analysis, fluorometric assays, SE-HPLC, RP-HPLC, as well as specific enzymatic tests.

Overview of Therapeutic Properties

Fabry's disease is a rare X-linked genetic disorder caused by a deficiency or total absence of α-galactosidase A, the lysosomal enzyme that cleaves α-galactose residues from glycosphingolipid Gb3 (also called ceramide trihexoside, CTH). The absence of the enzyme, or a low activity, leads to an accumulation of sphingolipids in various tissues and cells, including the kidneys, heart, skin, and nervous and gastrointestinal systems. The major symptoms are pain and various complications that lead to premature death when the patient is in his or her 40s or 50s. Enzyme replacement therapy is indicated to overcome the deficiency in enzymatic activity. The recombinant protein, with the natural glycosylation pattern, is recognized by specific receptors (recognizing mannose-6-phosphate) on the surface of the cell and directed to the lysosomes, where it is required.

Replagal should be administered every 2 weeks by intravenous administration at a dose of 0.2mg/kg. Clinical trials have been carried out on a very limited number of patients, due to the rarity of the disease (there are an estimated 500 to 1000 sufferers within the E.U.). Nevertheless, administration of Replagal showed a reduction in pain and pain medicaments in most of the patients and a reduction of the levels of accumulation of sphingolipids in tissues and in plasma and urine.

Mild allergic reactions related to the infusion were observed in 10% of the patients and decreased over time. Antibodies against the protein were observed in 55% of the patients. These antibodies did not seem to interfere with the safety and efficacy of the product, and the antibody titre decreased over time. The use of Replagal before appearance of disease symptoms should be considered, because the organ damage is irreversible. Replagal should not be administered with chloroquine, amiodarone, benoquin, or gentamicin.

Further Reading

http://www.eudra.org
http://www.tktx.com
Beck, M., Agalsidase alfa — a preparation for enzyme replacement therapy in Anderson-Fabry's disease, *Expert Opin. Invest. Drugs,* 11, 851–858, 2002.
Metha, A., Agalsidase alfa: specific treatment for Fabry's disease, *Hosp. Med.,* 63, 347–350, 2002.

Revasc

Product Name: Revasc (trade name)

Desirudin (international nonproprietary name)

Description: Revasc is a recombinant analogue of the anticoagulant hirudin, found in the saliva of the leech *Hirudo medicinalis*. It is similar to the natural molecule, but the tyrosine residue at position 63 lacks a sulphate group. Revasc is produced by recombinant DNA technology in modified yeast cells. The product is presented in a lyophilized form to be reconstituted using the solvent provided before subcutaneous administration.

Approval Date: 1997 (E.U.)

Therapeutic Indications: Prevention of deep venous thrombosis in patients undergoing elective hip and knee replacement surgery.

Manufacturer: Laboratories Ciba-Geigy S.A., 26 rue de la Chapelle BP 349, F-68330, Huninge Cedex, France, http//www.ciba.com

Marketing: Aventis Pharma S.A., 20 avenue Raymond Aron, 92165 Antony Cedex, France, http://www.aventis.com

Manufacturing

Revasc is a recombinant form of the natural hirudin found in the saliva of the leech *H. medicinalis*. It is produced by recombinant DNA technology in a modified *Saccharomyces cerevisiae* strain. It is very similar to the natural protein, differing only in a tyrosine at position 63, which lacks the sulphate group of the natural molecule. This modification does not affect the molecule's anticoagulant activity. During production in the host yeast cells, the protein is secreted into the culture medium. Purification involves a number of high-resolution chromatographic steps and the purified product is lyophilized subsequent to excipient addition and filter sterilization. The final formulation includes desirudin (active) as well as magnesium chloride and sodium chloride. The product is provided with a solvent, containing mannitol, for reconstitution before injection.

The shelf life of the product is 24 months when stored at less than 25°C and protected from light. Tests carried out to ensure the quality and safety of the product include SDS-PAGE, amino acid sequencing and peptide mapping, and tests for the presence of microbial and viral contaminants.

Overview of Therapeutic Properties

Patients undergoing orthopedic surgery may suffer from deep vein thrombosis and therefore are usually given anticoagulants, such as heparin. Revasc offers a number of advantages over other anticoagulants: it inactivates bound as well as free thrombin, it is not dependent upon antithrombin III, it has no effect on platelets, and it cannot be inactivated by antiheparin proteins.

A 15-mg dose of Revasc should be administered 5 to 15 minutes before surgery and then twice a day for up to 9 to 12 days afterward. Revasc proved to be more effective than heparin in preventing thrombotic events and pulmonary embolism, thus leading to increased survival rates. Minor bleedings appeared to be the most common side effect, but while they appeared more frequently and more profusely than after administration of heparin, they rarely led to major complications. No studies have been carried out on children. Revasc should not be administered in the case of hemophilia or during pregnancy and lactation.

Further Reading

http://www.aventis.com
http://www.eudra.org
Matheson, A.J. and Goa, K.L., Desirudin: a review of its use in the management of thrombotic disorders, *Drugs*, 60, 679–700, 2000.

Roferon A

Product Name:	Roferon A (trade name) Interferon alfa-2a (international nonproprietary name)
Description:	Roferon A is a recombinant human interferon alfa-2a. The 165-amino acid, 19-kDa single chain polypeptide is produced in *Escherichia coli* using recombinant DNA technology. It differs from the natural molecule in containing an additional methionine residue at its N terminus and in lacking glycosylation, though it retains the biological activity of the naturally occurring molecule. It is supplied as a liquid formulation in vials or prefilled syringes at various strengths (usually 6 to 36 million IU active/ml) for intravenous or subcutaneous administration.
Approval Date:	1986 for hairy cell leukemia, 1995 for chronic myelogenous leukemia, 1996 for chronic hepatitis C, 1998 for AIDS-associated Kaposi's sarcoma (U.S.)
Therapeutic Indications:	Roferon A is indicated for the treatment of adult patients diagnosed with chronic hepatitis C and hairy cell leukemia, Philadelphia chromosome-positive patients with chronic myelogenous leukemia in chronic phase, and adult patients with AIDS-associated Kaposi's sarcoma.
Manufacturer:	Hoffman-La Roche, Inc., 340 Kingsland Street, Nutley, NJ 07110-1199, http://www.rocheusa.com
Marketing:	Hoffman-La Roche, Inc., 340 Kingsland Street, Nutley, NJ 07110-1199, http://www.rocheusa.com

Manufacturing

The recombinant form of human interferon alfa-2a is produced in *E. coli* and differs from the natural molecule in having an additional methionine residue at its N-terminal end and in lacking glycosylation. The recombinant protein is purified from the bacterial culture using several chromatographic steps including affinity, ion exchange, and size exclusion chromatography. It is presented as a liquid formulation containing interferon alfa-2a, sodium chloride, polysorbate 80, benzyl alcohol, and ammonium acetate. Roferon A is supplied in single- and multiuse vials, as well as in prefilled syringes, for intramuscular or subcutaneous administration.

Validation tests, which ensure the quality and safety of the product, include SDS-PAGE, amino acid composition, tryptic digest mapping, analysis of DNA content, and antiviral assays.

Overview of Therapeutic Properties

Roferon A exhibits the antiviral and antitumor activity associated with naturally occurring human interferon alfa (i.e., inhibition of viral replication, prevention of proliferation of tumor cells, and modulation of the immune response). Roferon A is indicated for the treatment of adult patients with chronic hepatitis C, hairy cell leukemia, AIDS-related Kaposi's sarcoma, and for Philadelphia chromosome-positive patients with chronic myelogenous leukemia. It is administered three times a week for 12 months to chronic hepatitis C patients who have been diagnosed with liver biopsy and viral immunodetection tests. The appropriate induction dose is administered daily for 16 to 24 weeks in patients with hairy cell leukemia, followed by a maintenance dose three times a week. Patients with AIDS-related Kaposi's sarcoma are assessed for their response likelihood prior to commencing therapy. The induction dose is administered for 10 to 12 weeks, followed by a maintenance dose three times a week. Ph-positive patients with chronic myelogenous leukemia receive a daily dose, though the optimal duration of the treatment has yet to be established; the median time for complete hematological response was 5 months, but a response was observed for up to 18 months after commencement of the treatment. Clinical trials showed that Roferon A induced a response in patients with chronic hepatitis C, leading to an improvement in liver histology and a reduction in the viral titre. It induced disease stabilization and tumor regression in patients with hairy cell leukemia and Kaposi's sarcoma, and it reduced disease progression and increased the overall survival of patients with chronic myelogenous leukemia, compared to chemotherapy.

The most commonly reported side effects are flu-like symptoms, headaches, injection site reactions, nausea, and vomiting. Depression and suicidal behavior, as well as fatal ischemic and autoimmune and infectious disorders, have also been observed. Roferon A is contraindicated during pregnancy and lactation.

Further Reading

http://www.fda.gov
http://www.rocheusa.com
Haria, M. and Benfield, P., Interferon -alpha-2a. A review of its pharmacological properties and therapeutic use in the management of viral hepatitis, *Drugs*, 50, 873–896, 1995.
Williams, C.D. and Linch, D.C., Interferon alfa-2a, *Br. J. Hosp. Med.*, 57, 436–439, 1997.

Saizen

Product Name: Saizen (trade name)

Somatropin (rDNA) for injection (common name)

Description: Saizen is a human growth hormone (hGH) produced by recombinant DNA technology in an engineered mammalian cell line. The 191-amino acid, 22.125-kDa protein displays an amino acid sequence identical to the pituitary-derived endogenous growth hormone. The product is presented in freeze-dried format and is generally available in vials containing 5.0 or 8.8 mg active ingredient. The diluent provided is bacteriostatic water for injection (WFI containing 0.9% benzyl alcohol). After reconstitution, the product, which displays a pH of 6.5 to 8.5, is administered subcutaneously or intramuscularly. The active ingredient in Saizen is identical to that in Serono's other hGH product, Serostim.

Approval Date: 1996 (U.S.)

Therapeutic Indications: Saizen is indicated for the long-term treatment of children with growth failure due to inadequate secretion of endogenous growth hormone.

Manufacturer: Serono Inc., One Technology Place, Rockland, MA 02370, http://www. serono.usa.com (U.S.)

Marketing: Serono Inc., One Technology Place, Rockland, MA 02370, http://www. serono.usa.com (U.S.)

Manufacturing

Saizen is synthesized in a mammalian cell line (murine tumor cell line C-127) that has been engineered to carry the hGH gene. Upstream processing involves culture of the producer cell line in an animal cell bioreactor. Biologically active recombinant hGH is secreted by the producer cells directly into the extracellular media. After removal of the cell mass, the hormone is subject to downstream processing. This includes various concentration, filtration, and multiple chromatographic (purification) steps. Sucrose and O-phosphoric acid are added as excipients, with further pH adjustment using O-phosphoric acid or sodium hydroxide, as appropriate. The product is then sterile filtered, aseptically filled into presterile vials, and lyophilized.

Overview of Therapeutic Properties

Preclinical and clinical studies demonstrate that Siazen is biologically and therapeutically equivalent to endogenous pituitary-derived hGH. Saizen has been shown to stimulate skeletal growth in hGH-deficient prepubertal children. This effect is promoted by hGH and its main biological mediator, insulin-like growth factor-1 (IGF-1). The product also promotes organ growth and influences carbohydrate, lipid, mineral, and connective tissue metabolism in the same way as the native hormone.

Bioavailability of Saizen after subcutaneous administration ranges between 70 and 90%, and the mean half-life of the product after subcutaneous or intramuscular administration is 1.75 and 3.4 hours, respectively. Saizen dosage and administration schedules can be tailored to meet individual patient needs, but a dosage of 0.06 mg/kg administered three times weekly is recommended. Treatment with Saizen should be discontinued when the epiphyses are fused.

Various adverse events associated with Saizen administration are only infrequently witnessed. They include negative local reactions at the sites of injection, hypothyroidism, hypoglycemia, seizures, and exacerbation of preexisting psoriasis. Leukemia has been reported in a small number of children treated with hGH, although a direct link with the product has not been established. Saizen is contraindicated in the presence of active neoplasia, and treatment should not be initiated in patients with acute critical illness, particularly those suffering from acute respiratory failure or complications due to heart or abdominal surgery.

Further Reading

Bercu, B.B. et al., Long-term therapy with recombinant human growth hormone (Saizen) in children with idiopathic and organic growth hormone deficiency, *Endocrine*, 15(1), 43–49, 2001.

Keller, E. et al., Use of manufactured Saizen growth hormone in treatment of patients for hypothalamico-hypophyseal hyposomia, *Z. Klin. Med.*, 45(17), 1509–1512, 1990.

Murray, F.T. et al., Prolonged growth response to Saizen in pediatric subjects with growth hormone deficiency who responeded poorly to GHRH (Geref) therapy, *Pediatr. Res.*, 45(4), 546, 1999.

Serostim

Product Name: Serostim (trade name)

Somatropin (rDNA) for injection (common name)

Description: Serostim is a human growth hormone (hGH) produced by recombinant DNA technology in an engineered mammalian cell line. The 191-amino acid, 22.125-kDa protein displays an amino acid sequence identical to the pituitary-derived endogenous growth hormone. The product is presented in freeze-dried format and is generally available in single-dose vials containing 4.0, 5.0, or 6.0 mg active ingredient. The diluent provided is water for injection (WFI). After reconstitution, the product, which displays a pH of 7.4 to 8.5, is administered subcutaneously. The active ingredient in Serostim is identical to that in Serono's other hGH product, Saizen.

Approval Date: 1996 (U.S.)

Therapeutic Indications: Serostim is indicated for the treatment of HIV patients with wasting or cachexia to increase lean body mass and body weight and improve physical endurance. Concomitant antiretroviral therapy is necessary.

Manufacturer: Serono Inc., One Technology Place, Rockland, MA 02370, http://www. serono.usa.com (U.S.)

Marketing: Serono Inc., One Technology Place, Rockland, MA 02370, http://www. serono.usa.com (U.S.)

Manufacturing

Serostim is synthesized in a mammalian cell line (murine tumor cell line C-127) that has been engineered to carry the hGH gene. Upstream processing involves culture of the producer cell line in an animal cell bioreactor. Biologically active recombinant hGH is secreted by the producer cells directly into the extracellular media. After removal of the cell mass, the hormone is subject to downstream processing. This includes various concentration, filtration, and multiple chromatographic (purification) steps. Sucrose and O-phosphoric acid are added as excipients, with further pH adjustment using O-phosphoric acid or sodium hydroxide, as appropriate. The product is then sterile filtered, aseptically filled into presterile vials, and lyophilized.

Overview of Therapeutic Properties

Serostim is biologically and therapeutically equivalent to endogenous pituitary-derived hGH. HIV-associated wasting or cachexia is characterized by loss of lean body mass and body weight. Lean body mass includes skeletal muscle, organ tissue, blood, and intracellular and extracellular water. Depletion of lean body mass results in muscle weakness, organ failure, and eventually death. Nutritional intervention on its own generally results in supplemental calories being converted mainly into body fat. Trials have shown that concurrent administration of Serostim resulted in not only further weight gain (1.6 kg over 12 weeks) but increased lean body mass (3.6 kg over 12 weeks). Treatment with Serostim was also found to significantly increase physical function as measured by treadmill exercise testing. Clinical studies with HIV patients found that Serostim also improved nitrogen balance and resulted in enhanced retention of phosphorous, potassium, sodium, and nitrogen.

Bioavailability of Serostim after subcutaneous administration ranges between 70 and 90%, and the mean half-life of the product after subcutaneous administration is 3.9 hours. Serostim dosage and administration schedules can be tailored to meet individual patient needs, but the usual starting dose is daily subcutaneous administration of 0.1 mg/kg. Reduced side effects are noted if the product is administered every second day. Most of the effects are fully attained within 12 weeks of treatment initiation.

Various adverse events may be associated with administration of Serostim. The most common adverse reactions were muscoskeletal discomfort and increased tissue turgor (swelling, particularly of the hands and feet). Additional side effects sometimes noted included hyperglycemia, gastrointestinal disorders, respiratory symptoms, headaches, and fatigue.

Serostim is contraindicated in the presence of active neoplasia, and treatment should not be initiated in patients with acute critical illness, particularly those suffering from acute respiratory failure or complications due to heart or abdominal surgery. Under some conditions, hGH has been shown to potentiate HIV replication *in vitro*. Patients therefore are maintained on antiretroviral drugs for the duration of hormone treatment.

Further Reading

Bristow, A.F. and Jespersen, A.M., The second international standard for somatropin (recombinant DNA-derived human growth hormone): preparation and calibration in an international collaborative study, *Biologicals*, 29 (2), 97–106, 2001.

Pettit, R.D. et al., The use of Serostim (recombinant human growth hormone) in mild to moderate HIV wasting in the post HAART era, *Clin. Infect. Dis.*, 31 (1), 380, 2000.

Scheperle, M. et al., A case management approach for two diabetic patients receiving HAART and recombinant human growth hormone (Serostim) for HIV-associated wasting, *AIDS*, 14, 146, 2000.

Simulect

Product Name: Simulect (trade name)

Basiliximab (international nonproprietary name)

Description: Simulect is a chimeric (murine-human) monoclonal antibody with specific binding activity for the α-chain (CD25) of the human interleukin-2 (IL-2) receptor, which is expressed on the surface of activated lymphocytes. It is produced by recombinant DNA technology in murine myeloma cell culture and is supplied in a lyophilized form to be reconstituted for infusion or intravenous injection.

Approval Date: 1998 for prophylaxis of acute organ rejection in patient receiving renal transplantation (E.U. and U.S.); 2001 for pediatric use and for administration as an intravenous bolus injection (U.S.)

Therapeutic Indications: Simulect is indicated for the prophylaxis of acute kidney rejection in *de novo* allogeneic transplantation as part of an immunosuppressive regimen including cyclosporine and corticosteroids.

Manufacturer: Novartis Pharma S.A., 68330 Huningue, France, http://www.novartis.com (manufacturer is responsible for import and batch release in the European Economic Area)

Novartis Pharmaceutical Corporation, 59 Route 10, East Hanover, NJ 07936-1080, http://www.pharma.us.novartis.com (U.S.)

Marketing: Novartis Europharm Limited, Wimblehurst Road, Horsham, West Sussex, RH12 5AB, U.K., http://www.novartis.com (E.U.)

Novartis Pharmaceutical Corporation, 59 Route 10, East Hanover, NJ 07936-1080, http://www.pharma.us.novartis.com (U.S.)

Manufacturing

The chimeric antibody Simulect was constructed by recombinant DNA technology, combining the variable regions of the murine monoclonal antibody RFT5, raised against the α-chain of the human IL-2 receptor and constant domains from a human IgG1. The recombinant antibody is produced in cell culture of murine myeloma cells. The purification process comprises eight steps, including chromatographic steps and procedures for removal and inactivation of viruses. The final product, extensively characterized, is presented in a lyophilized form consisting of basiliximab (active), potassium dihydrogen phosphate, disodium hydrogen phosphate, sodium chloride, sucrose, mannitol, and glycine (as excipients).

The shelf life of the product is 12 months when stored at 2 to 8°C. Extensive control tests are carried out during processing and on the final product to ensure the quality and safety of the product.

Overview of Therapeutic Properties

The recombinant antibody Simulect binds specifically to the α-chain of the IL-2 receptor on the surface of lymphocytes. This prevents the binding of IL-2, in turn leading to inhibition of IL-2 mediated proliferation of T cells. Suppression of the immunoresponse minimizes the occurrence of organ rejection. Simulect, unlike other immunosuppressive drugs, acts mainly on activated lymphocytes expressing a high level of IL-2 receptor.

Simulect is administered to patients who receive renal transplantation as part of an immunoreppressive regimen, usually consisting of cyclosporine and corticosteroids (double therapy) or cyclosporine, corticosteroids, and azathioprine (triple therapy). It is administered as an infusion over a period of 20 to 30 minutes or injected directly using a syringe. A first dose is given 2 hours before the transplantation, and a second dose is given 4 days later. A total of 40 mg is administered to adults, while children (less than 35 kg) receive a total amount of 20 mg. IL-2 receptor is kept completely blocked for 4 to 6 weeks, the period during which most rejections occur.

Pivotal studies on the use of Simulect as part of an immunosuppressant regimen showed a significantly reduced incidence of acute rejection episodes at 6 and 12 months after transplantation, while no effect was found on graft survival. In patients treated with Simulect, most of the graft losses were not due to rejection and late rejection did not occur. The use of Simulect did not increase side effects when compared to patients receiving placebos. Human antihuman antibody and human antichimeric antibody responses were observed rarely. Severe acute hypersensitivity reactions have been observed. Simulect is contraindicated during pregnancy and lactation.

Further Reading

http://www.eudra.org
http://www.fda.gov
http://www.novartis.com
Nashan, B. et al., Randomised trial of basiliximab versus placebo for control of acute cellular rejection in renal allograft recipients. CHIB 201 International Study Group, *Lancet*, 350, 1193–1198, 1997.
Onrust, S.V. and Wiseman, L.R., Basiliximab, *Drugs*, 57, 207–214, 1999.
Ponticelli, C. et al., A randomized, double-blind trial of basiliximab immunoprophylaxis plus triple therapy in kidney transplant recipients 1,2, *Transplantation*, 72, 1261–1267, 2001.

Somavert

Product Name: Somavert (trade name)

Pegvisomant (international nonproprietary name)

Description: Pegvisomant, the active ingredient of Somavert, is a PEGylated recombinant analogue of human growth hormone (hGH). It has been engineered to introduce nine mutations into the hGH amino acid sequence. This analogue binds to the hGH cell surface receptor but does not trigger an intracellular response. As such, it effectively acts in an antagonistic fashion, reducing the effects of endogenous hGH. The molecule is PEGylated *in vitro* in order to increase its serum half-life. It is generally presented in lyophilized form, to be reconstituted prior to subcutaneous use.

Approval Date: 2002 (E.U.), 2003 (U.S.)

Therapeutic Indications: Somavert is indicated for the treatment of patients with acromegaly who have inadequate responses to surgery or radiation therapy and in whom appropriate medical treatments with somatostatin analogues did not normalize IGF-1 concentrations or were not tolerated.

Manufacturer: Pharmacia N.V./A.S., Rijksweg 12, B-2870 Puurs, Belgium (manufacturer is responsible for import and batch release in European Economic Area)
Abbott Laboratories, 1401 Sheridan Road, North Chicago, IL 60064 (U.S.), http://www.abbott.com

Marketing: Pharmacia Enterprises S.A., 6, Circuit de la Foire Internationale, BP 2507, L-1025, Luxembourg, G.D., Luxembourg (E.U.)
Pharmacia and Upjohn Company, 7000 Portage Road, Kalamazoo, MI 49001 (U.S.)

Manufacturing

The 191-amino acid hGH analogue is produced in an engineered strain of *Escherichia coli.* The unpegylated protein is extracted from the producer cells subsequent to fermentation. Purification entails a series of chromatographic and ultrafiltration steps. The protein is then PEGylated (covalent attachment of activated polyethylene glycol molecules). Each Pegvisomant molecule contains four to five individual PEG molecules. The active ingredient is then formulated with the following excipients: mannitol, glycine, sodium phosphate dibasic, and monobasic. After sterilization by filtration, the product is aseptically filled into sterile glass vials and lyophilized. The solvent for reconstitution (water for injection) is provided in a separate vial. The product has been assigned a shelf life of 9 months when stored at less than 25°C, protected from sunlight.

Overview of Therapeutic Properties

Acromegaly is a rare endocrine disorder characterized by elevated plasma hGH levels. This is generally caused by a benign pituitary adenoma (tumor) triggered by a gene mutation. The condition most commonly effects middle-aged adults. Patients display significantly increased morbidity and mortality and usually suffer from cardiovascular, cerebrovascular, and respiratory diseases. Increased incidences of malignancies and lung infections are also characteristic. Additional complications related to the condition include acral enlargement, disfigurement, and hypertension. The clinical manifestations appear to be mediated primarily by (hGH-triggered) elevated serum insulin-like growth factor-1 (IGF-1) concentrations.

First choice therapy for acromegaly is usually surgical excision of the tumor. Radiotherapy on its own or in conjunction with surgery is also commonly employed. An alternative therapeutic route entails administration of somatostatin analogues, leading to inhibition of growth hormone secretion and, consequently, a reduction in serum IGF-1 levels.

Pegvisomant binds to endogenous growth hormone receptors but fails to trigger an intracellular response. It can displace and prevent binding of endogenous GH to its receptor, thereby effectively reducing the biological effects of excessive circulatory hGH levels. Clinical trials reveal that administration of Pegvisomant normalizes serum IGF-1 levels within weeks of commencement of therapy. The most common adverse effects included injection site reactions, sweating, headache, and asthenia.

Further Reading

http://www.eudra.org

Anon., Pegvisomant — Trovert — Somavert — treatment of acromegaly, *Drug Future,* 26(9), 911–913, 2001.

Clark, R., Olson, K., Fiuh, G. et al., Long-acting growth hormones produced by conjugation with polyethylene glycol, *J. Biol. Chem.*, 271(36), 21969–21977, 1996.

Fassanacht, M. et al., Octreotide LAR treatment throughout pregnancy in an acromegalic women, *Clin. Endocrinol. (Oxford),* 55(3), 411–415, 2001.

Parkinson, C. and Trainer, P.J., The place of Pegvisomant in the management of acromegaly, *Endocrinologist,* 13(5), 408–416, 2003.

Synagis

Product Name:	Synagis (trade name) Palivizumab (international nonproprietary name)
Description:	Synagis is a humanized monoclonal antibody that recognizes and binds the A antigenic site of the fusion protein of the respiratory syncytial virus (RSV). It is produced by recombinant DNA technology in a myeloma NSO cell line and is supplied in a lyophilized form to be reconstituted for intramuscular administration.
Approval Date:	1998 (U.S.); 1999 (E.U.)
Therapeutic Indications:	Synagis is indicated for the prevention of serious lower respiratory tract disease requiring hospitalization and caused by RSV in high risk RSV disease patients, such as premature infants and children with bronchopulmonary dysplasia.
Manufacturer:	Abbott S.p.A., 04010 Campoverde di Aprilia (Latina), Italy, http://www.abbott.com (manufacturer is responsible for import and batch release in the European Economic Area) MedImmune, Inc., 35 West Watkins Mill Road, Gaithersburg, MD 20878, http://www.medimmune.com (U.S.)
Marketing:	Abbott Laboratories Limited, Queenborough, Kent ME11 5EL, U.K., http://www.abbott.com (E.U.) MedImmune, Inc., 35 West Watkins Mill Road, Gaithersburg, MD 20878, http://www.medimmune.com (U.S.)

Manufacturing

The humanized monoclonal antibody (95% human, 5% murine in sequence) was constructed by combining the complementarity determining regions

(CDRs) of the murine monoclonal antibody Mab 1129 with human IgG1 constant regions. The recombinant antibody binds specifically to the antigenic site A of the fusion protein of RSV, neutralizing the virus.

The murine myeloma NSO cell line is used to produce the recombinant antibody in a bioreactor. The purification process includes microfiltration to remove myeloma cells and debris, three-step chromatography, and several viral removal steps. The final product is presented in a lyophilized form consisting of palivizumab (active), as well as glycine, histidine, and mannitol as excipients.

The shelf life of the product is 24 months when stored at 2 to 8°C. Routine tests are carried out during processing and on the final product to evaluate potency, identity, presence of endotoxins, mycoplasma, and viral and bacterial safety.

Overview of Therapeutic Properties

RSV is a common virus that causes most of the severe lower respiratory tract illness in children worldwide. Hospitalization is required in particularly severe cases, mostly affecting high risk patients such as premature infants and children with bronchopulmonary dysplasia or congenital heart disease. The recombinant antibody Synagis neutralizes RSV by binding to the viral fusion protein and by preventing the formation of complexes of fused RSV-infected cells (syncytia). The epitope recognized by the antibody is antigenicaly stable over time and geographically and is present in both groups of RSV subtypes A and B.

Synagis should be administered monthly to high-risk patients prior to the beginning of and throughout the RSV disease season. Studies show that Synagis reduced the hospitalization rate of patients with RSV disease by 55%, meaning that 1 hospitalization was avoided in every 17 patients treated.

Synagis is safe and well tolerated and has rare and mild side effects and a human antihuman antibody response (HAHA) of no clinical relevance. Administration of Synagis does not affect the severity of the RSV disease, unlike formalin-inactivated RSV vaccine. Synagis was shown to be more potent in neutralizing RSV than was RespiGam, a RSV hyperimmune globulin, and has the advantage over RespiGam of intramuscular administration (whereas RespiGam requires intravenous administration). Synagis was also shown to be more potent against RSV than was the humanized antibody RSHZ19, which was discontinued after pivotal trial.

Further Reading

http://www.eudra.org
http://www.fda.gov
http://www.medimmune.com
Saez-Llorens, X. et al., Safety and pharmacokinetics of an intramuscular humanized monoclonal antibody to respiratory syncytial virus in premature infants and infants with bronchopulmonary dysplasia. The MEDI-493 Study Group, *Pediatr. Infect. Dis. J.*, 17, 787–791, 1998.
Scott, L.J. and Lamb, H.M., Palivizumab, *Drugs*, 58, 305–313, 1999.
Sorrentino, M. and Powers, T., Effectiveness of palivizumab: evaluation of outcomes from the 1998 to 1999 respiratory syncytial virus season. The Palivizumab Outcomes Study Group, *Pediatr. Infect. Dis. J.*, 19, 1068–1071, 2000.

Tecnemab K1 (withdrawn from market)

Product Name: Tecnemab K1 (trade name)

Antimelanoma Mab fragments (international nonproprietary name)

Description: The active substance of Tecnemab K1 is a mixture of monovalent and bivalent antibody fragments [F(ab) and F $(ab)_2$ fragments] derived from a murine IgG 2a monoclonal antibody raised against high molecular weight melanoma associated antigen (HMW-MAA). HMW-MAA is a surface antigen associated with melanoma lesions. The product is provided as a kit containing 3 x 6 ml vials, each containing 0.35 mg of anti-HMW-MAA monoclonal antibody fragments. Additional ingredients include stannous chloride, human serum albumin, sodium tartrate, and potassium biphthalate. Also included in the kit are 3 x 1 ml ion exchange columns. When used, the antibody fragments are labeled with the ^{99m}Tc radioisotope, which is not supplied with the kit. Immediately before use, the separately purchased ^{99m}Tc is coupled to the antibody fragments, followed by separation of free from bound radioactivity using the ion exchange columns.

Approval Date: 1996 (E.U.)

Withdrawal Date: This product no longer appears to be on the market. Although not listed on the European Medicines Evaluation Agency's (EMEA's) list of product withdrawals, its European public assessment report is no longer available on the EMEA Web site, and it is not listed on the European Commission's register of approved medicinal products.

Therapeutic Indications: This product was recommended as an adjunct to other diagnostic procedures for visualization by radioimmunoscintigraphy (RIS) of regional lymph node and distant metastases in staging and follow-up of patients with stage I-III melanoma. It assisted in differential diagnosis of suspected ocular melanoma.

Manufacturer: Sorin Biomedica Diagnostics S.p.A., Italy (manufacturer was responsible for batch release in the European Economic Area)

Marketing: Sorin Biomedica Diagnostics S.p.A., Italy

Manufacturing

Intact antibody is produced by culture of the appropriate hybridoma cell line in the peritoneal cavity of mice followed by collection of the antibody-containing ascitic fluid. Subsequent purification involves protein A-based affinity chromatography and ion exchange chromatography. The antibody is fragmented enzymatically.

Overview of Therapeutic Properties

Clinical studies illustrated that RIS imaging after administration of ^{99m}Tc-labeled product provided a specific and relatively sensitive diagnostic tool for imaging melanoma relapses. No serious adverse events were reported during initial trials.

Thyrogen

Product Name: Thyrogen (trade name)

Thyrotropin alfa (international nonproprietary name)

Description: Thyrogen is a recombinant human thyroid-stimulating hormone (TSH) produced in Chinese hamster ovary (CHO) cells using recombinant DNA techniques. The active hormone is a hetrodimer, consisting of a 92-amino acid α-subunit and a 118-amino acid β-subunit held together noncovalently. Both subunits are glycosylated. The recombinant product consists of a number of different isoforms, which slightly differ from the natural protein due to differences in post-translational processing. Nevertheless, activity comparable to the natural hormone is retained. It binds TSH receptors and stimulates iodine uptake and the production of thyroid hormones. Thyrogen is presented as a lyophilized powder (0.9 mg active/vial) to be reconstituted before intramuscular administration.

Approval Date: 1998 (U.S.); 2000 (E.U.)

Therapeutic Indications: Thyrogen is indicated for radioiodine imaging and in combination with serum thyroglobulin testing for detection of thyroid cancer in patients after thyroid removal and throughout hormone suppression therapy.

Manufacturer: Genzyme Ltd., 37 Hollands Road, Haverhill, Suffolk CB9 8PU, U.K., http://www.genzyme.com (manufacturer is responsible for batch release in the European Economic Area)
Genzyme Corporation, Inc., One Kendall Square, Cambridge, MA 01242, http://www.genzyme.com (U.S.)

Marketing: Genzyme Europe B.V., Gooimeer 10, 1411 DD Naarden, the Netherlands, http://www.genzyme.com (E.U.)
Genzyme Corporation, Inc., One Kendall Square, Cambridge, MA 01242, http://www.genzyme.com (U.S.)

Manufacturing

Thyrogen is a recombinant TSH. It is produced in a modified CHO cell line using recombinant DNA technology. After cell culture, in which the TSH is produced as an extracellular product, the bioreactor harvest is filtered. This is followed by concentration of the crude product using ultrafiltration. The TSH is then extensively purified via several chromatographic steps and filtration procedures. The final product consists of the active substance thyrotropin alfa, as well as mannitol, monohydrate monobasic sodium phosphate, heptahydrate dibasic sodium phosphate, and sodium chloride as excipients.

The shelf life of the product is 36 months when stored at 2 to 8°C, protected from light. Extensive tests carried out to ensure the quality and the safety of the product include spectrophotometric analysis, ELISA, native PAGE, IEF, SEC-HPLC, bioassays to determine activity, and stringent analysis to detect virus and bacterial contaminants.

Overview of Therapeutic Properties

Thyrogen is indicated for radioiodine imaging for the detection of thyroid cancers in adults from whom the thyroid gland has been removed and who are still undergoing hormone suppression therapy. It is used in combination with serum thyroglobulin testing. Thyrogen offers the advantage that hormone suppression therapy can be continued throughout its use, while previous products required therapy to be stopped during their use. Thyrotropin alfa binds to TSH receptors and stimulates the uptake of iodine, as well as the production of thyroglobulin and thyroid hormones; its activity is comparable to the natural thyroid-stimulating hormone.

Two doses of 0.9 mg of thyrotropin alfa should be given intramuscularly, with a 24-hour interval between the two doses. After 24 hours, radioiodine should be administered, and the scanning for imaging should be carried out after a further 48 to 72 hours. Thyroglobulin testing should take place 72 hours after the administration of Thyrogen.

Studies showed that Thyrogen is equally effective upon suspension or throughout continued hormone suppression therapy. Reactions at the sites of injection, nausea, headaches, flu-like symptoms, and weakness were the most commonly reported side effects. Thyrogen is contraindicated during pregnancy and lactation.

Further Reading

http://www.eudra.org
http://www.fda.gov
http://www.genzyme.com
http://www.thyrogen.com

Mariani, G. et al., Clinical experience with recombinant human thyrotropin (rhTSH) in the management of patients with differentiated thyroid cancer, *Cancer Biother. Radiopharm.*, 15, 211–217, 2000.

Reiners C. et al., Clinical experience with recombinant human thyroid-stimulating hormone (rhTSH): whole-body scanning with iodine-131, *J. Endocrinol. Invest.*, 22(11 Suppl.), 17–24, 1999.

Triacelluvax (withdrawn from market)

Product Name: Triacelluvax (trade name)

Diphtheria, tetanus, and acellular pertussis vaccine (common name)

Description: Triacelluvax was a vaccine preparation against diphtheria, tetanus, and pertussis. It consists of diphtheria and tetanus toxoids, as well as three purified acellular pertussis antigens purified from cultures of a genetically engineered *Bordetella pertussis* strain. The product was formulated in a glass syringe as a suspension for injection. A single monodose preparation contained 30 IU diphtheria toxoid, 40 IU tetanus toxoid, 5 μg genetically modified pertussis toxin, 2.5 μg filamentous hemagglutinin (FHA) and 2.5 μg pertactin. The product was administered intramuscularly.

Approval Date: 1999 (E.U.)

Withdrawal Date: 2002 (for commercial reasons)

Therapeutic Indications: Triacelluvax was indicated for active immunization of children from 6 weeks to 7 years of age against diphtheria, tetanus, and pertussis.

Manufacturer: Chiron S.p.A., Via Fiorentina, 1, 53100, Siena, Italy

Marketing: Chiron S.p.A., Via Fiorentina, 1, 53100, Siena, Italy

Manufacturing

The diphtheria toxoid is obtained by fermentation of an appropriate strain of *Corynebacterium diphtheriae.* After fermentation, the cell mass is removed by filtration and formaldehyde is added to the extracellular media in order to chemically inactivate the diphtheria toxin. The formaldehyde-inactivated toxoid is then purified by depigmentation, centrifugation, filtration, and ultrafiltration. The tetanus toxoid is obtained by fermentation of an appropriate strain of *Clostridium tetani*, with recovery, inactivation, and purification of the toxoid being undertaken by means very similar to those already described for the diphtheria toxoid. The modified pertussis toxin, as well as FHA and pertactin, is obtained by fermentation of a genetically engineered strain of *B. pertussis*. Pertussis toxin is a 105-kDa protein composed of five subunits, one of which (the S1 subunit) mediates its toxic effects. Site-directed mutagenesis has been employed to make two amino acid substitutions in the S1 subunit, rendering the molecule nontoxic. The modified *B. pertussis* strain continues to express wild type FHA and pertactin. All three antigens are chromatographically purified following fermentation. The final product is obtained by formulating the active ingredients with sodium chloride, thiomersal, and aluminium hydroxide in water for injections.

Overview of Therapeutic Properties

Clinical trials showed that vaccine administration elicited an acceptable immune response to the various component antigens. Local and systematic reactions were less than those of the whole cell pertussis vaccines, studied for comparative purposes. The most frequently observed side effects included fever as well as swelling, redness, and tenderness at the site of injection.

Further Reading

http://www.eudra.org

Matheson, A. and Goa, K., Diphtheria-tetanus-acellular pertussis vaccine adsorbed (Triacelluvax; DtaP3-CB): a review of its use in the prevention of bordetella pertussis infection, *Pediatr. Drugs*, 2(2), 139–159, 2000.

Salmaso, S., Mastrantonio, P., et al., Sustained efficacy during the first 6 years of life of 3 component acellular pertussis vaccines administered in infancy: the Italian experience, *Pediatrics*, 108(5), E 81, 2001.

Tritanrix HepB

Product Name: Tritanrix HepB (trade name)

Combined vaccine DTP_w-HB (international nonproprietary name)

Description: Tritanrix HepB is a combined vaccine against diphtheria, tetanus, pertussis, and hepatitis B. It is a combination of two previously marketed vaccines: DTP_w, containing diphtheria toxoid, tetanus toxoid, and pertussis inactivated strain; and Engerix-B, a recombinant vaccine containing the major surface antigen of the hepatitis B virus. Tritanrix HepB is presented as a suspension for intramuscular injection.

Approval Date: 1996 (E.U.)

Therapeutic Indications: Tritanrix HepB is indicated for primary and booster vaccination against diphtheria, tetanus, pertussis, and hepatitis B in infants from 6 weeks of age.

Manufacturer: GlaxoSmithKline Biologicals, Rue de l'institut 89, 1330 Rixensart, Belgium, http://www.gsk.com (manufacturer is responsible for batch release in the European Economic Area)

Marketing: GlaxoSmithKline Biologicals, Rue de l'institut 89, 1330 Rixensart, Belgium, http://www.gsk.com

Manufacturing

Tritanrix HepB is a combination of two vaccines already on the market, DTP_w and Engerix-B. Diphtheria and tetanus toxoids are obtained from cultures of *Corynebacterium diphtheria* and *Clostridium tetani*, respectively, and are inactivated by the use of formalin, purified and adsorbed on hydrated aluminium oxide. The pertussis component is derived via heat inactivation of the strain *Bordetella pertussis*, while the major surface antigen of the hepatitis B virus is produced by recombinant DNA technology in *Saccharomyces cerevisiae*. The pertussis component and the hepatitis B antigen are adsorbed on aluminium phosphate. The final product, Tritanrix HepB, contains the antigenic components, hydrated aluminium oxide (as an adjuvant), aluminium phosphate (as an adjuvant), phenoxyethanol (as a preservative), sodium chloride, and traces of thiomersal. Tritanrix HepB is supplied as a suspension for intramuscular administration.

The shelf life of the product is 36 months when stored at 2 to 8°C in a light-protected container. The purity, potency, and safety of the product are ensured with extensive testing.

Overview of Therapeutic Properties

Tritanrix HepB is indicated for primary and booster vaccination of infants against diphtheria, tetanus, pertussis, and hepatitis B. Tritanrix HepB should be administered intramuscularly as a three-dose vaccine with at least 1 month between each dose in infants older than 8 weeks old. In infants with a high risk of infection, the first dose can be administered at 6 weeks of age. A booster vaccination may be administered in the second year of life.

Tritanrix HepB was found to be as effective as separately administered vaccines in eliciting immune protection against diphtheria, tetanus, pertussis, and hepatitis B. Administration of Tritanrix HepB with oral antipolio and anti-*Haemophilus influenzae* type b vaccines did not appear to interfere with the efficacy of the immune protection.

The side effects most commonly observed after administration of the combined vaccine were mild and similar to those typically observed for other vaccines, such as reactions at the injection site, vomiting, diarrhea, fever, and irritability. Tritanrix HepB should not be administered in cases of fever.

Further Reading

http://www.eudra.org

http://www.gsk.com

Poovorawan, Y. et al., Comparison study of combined DTP_w-HB vaccines and separate administration of DTP_w and HB vaccines in Thai children, *Asian Pac. J. Allergy Immunol.*, 17, 113–120, 1999.

Prikazsky, V. and Bock, H.L., Higher antihepatitis B response with combined DTP_w-HBV vaccine compared with separate administration in healthy infants at 3, 4, and 5 months of age in Slovakia, *Int. J. Clin. Pract.*, 55, 156–161, 2001.

Twinrix Adult

Product Name:	Twinrix Adult (trade name) Combined hepatitis A and hepatitis B vaccine (international nonproprietary name)
Description:	Twinrix Adult is a combined vaccine consisting of two separately marketed vaccines: Havrix Adult, containing inactivated hepatitis A virus, and Engerix-B, a recombinant vaccine containing the purified hepatitis B surface antigen (rHBsAg). Twinrix Adult is presented as a suspension in vials or in prefilled syringes for intramuscular administration. Each dose contains 720 ELISA units of inactivated hepatitis A and 20 μg rHBsAg in a final volume of 1.0 ml.
Approval Date:	1996 (E.U.); 2001 (U.S.)
Therapeutic Indications:	Twinrix Adult is indicated for immunization against hepatitis A and hepatitis B in adults.
Manufacturer:	GlaxoSmithKline Biologicals, Rue de l'institut 89, 1330 Rixensart, Belgium, http://www.gsk.com (manufacturer is responsible for batch release in the European Economic Area) Glaxo SmithKline Biologicals, Rue de l'institut 89, 1330 Rixensart, Belgium, http://www.gsk.com (U.S.)
Marketing:	GlaxoSmithKline Biologicals, Rue de l'institut 89, 1330 Rixensart, Belgium, http://www.gsk.com (E.U.) Glaxo SmithKline, 5 Moore Drive, Research Triangle Park, NC 27709, http://www.gsk.com (U.S.)

Manufacturing

Twinrix Adult is a combined vaccine consisting of the separately marketed vaccines Havrix and Engerix-B. Havrix contains the purified inactivated hepatitis A virus, derived from human diploid MRC-5 cells infected with the hepatitis A virus, while Engerix-B is a recombinant vaccine produced in

Saccharomyces cerevisiae, encoding the purified hepatitis B surface antigen (S protein). The inactivated hepatitis A virus is adsorbed on aluminium oxide, while the purified S protein from the hepatitis B virus is adsorbed on aluminium phosphate. Twinrix is then produced by simple combination of the bulk substances of the individual vaccines. The final formulation of Twinrix Adult contains the inactivated hepatitis A virus, S protein from the hepatitis B virus, hydrated aluminium oxide, aluminium phosphate, formaldehyde, neomycin sulphate, phenoxyethanol, sodium chloride, and traces of thiomersal. The product is provided as a suspension for intramuscular administration in vials or in prefilled syringes.

The shelf life of the product is 36 months when stored at 2 to 8°C in a light-protected container. The quality and the safety of the product are ensured by extensive testing.

Overview of Therapeutic Properties

Twinrix Adult is a combination of Havrix and Engerix-B, existing vaccines against the hepatitis A and hepatitis B virus, respectively. The product is indicated for the immunization of adult patients at risk of infection. Twinrix Adult should be administered by three intramuscular injections, with a 1-month interval between the first and second and 5 months between the second and third injections. When rapid protection against the hepatitis A and B viruses is required, three doses of Twinrix Adult may be administered within 1 month, a second dose after 1 week from the first, and the third dose after 2 weeks from the second dose. A subsequent single dose may be required after 6 months. The need for a booster administration has not been established.

The combined vaccine was shown to elicit an immune response within a few weeks from the first administration. Protection against hepatitis A and B was provided to the same extent as in the case of separately administered vaccines. The most common side effects were reactions at the site of injection, fatigue, nausea, headaches, and malaise. Caution should be taken with the administration of Twinrix Adult during pregnancy and lactation.

Further Reading

http://www.eudra.org
http://www.fda.gov
http://www.gsk.com
Joines, R.W. et al., A prospective, randomized, comparative US trial of a combination hepatitis A and B vaccine (Twinrix) with corresponding monovalent vaccines (Havrix and Engerix-B) in adults, *Vaccine*, 19, 4710–4719, 2001.

Nothdurf, H.D. et al., A new accelerated vaccination schedule for rapid protection against hepatitis A and B, *Vaccine,* 20, 1157–1162, 2002.
Thoelen, S. et al., The first combined vaccine against hepatitis A and B: an overview, *Vaccine,* 17, 1657–1662, 1999.

Twinrix Paediatric

Product Name: Twinrix Paediatric (trade name)

Combined hepatitis A and hepatitis B vaccine (international nonproprietary name)

Description: Twinrix Paediatric (identical in composition to the product Twinrix Adult, although administered in a different dose size) is a combined vaccine consisting of two separately marketed vaccines: Havrix Paediatric, containing inactivated hepatitis A virus, and Engerix-B, a recombinant vaccine containing purified hepatitis B surface antigen (rHBsAg). Twinrix Paediatric is presented as a suspension in vials or in prefilled syringes for intramuscular administration. Each dose contains 360 ELISA units of inactivated hepatitis A and 10 μg of rHBsAg in a final volume of 0.5 ml.

Approval Date: 1997 (E.U.)

Therapeutic Indications: Twinrix Paediatric is indicated for immunization against hepatitis A and hepatitis B in infants, children, and adolescents.

Manufacturer: GlaxoSmithKline Biologicals, Rue de l'institut 89, 1330 Rixensart, Belgium, http://www.gsk.com

Marketing: GlaxoSmithKline Biologicals, Rue de l'institut 89, 1330 Rixensart, Belgium, http://www.gsk.com

Manufacturing

Twinrix Paediatric is a combined vaccine consisting of the separately marketed vaccines Havrix and Engerix-B. Havrix contains the purified inactivated hepatitis A virus, derived from human diploid MRC-5 cells infected with the hepatitis A virus, while Engerix-B is a recombinant vaccine produced in *Saccharomyces cerevisiae* encoding the purified hepatitis B surface antigen (S protein). The inactivated hepatitis A virus is adsorbed on aluminium oxide, while the purified S protein from the hepatitis B virus is adsorbed on aluminium phosphate. Twinrix is then produced by simple combination of the bulk substances of the individual vaccines. The final formulation of Twinrix Paediatric contains the inactivated hepatitis A virus, S protein from hepatitis B virus, hydrated aluminum oxide, aluminium phosphate, formaldehyde, neomycin sulphate, phenoxyethanol, sodium chloride, and traces of thiomersal. The product is provided as a suspension for intramuscular administration in vials or in prefilled syringes.

The shelf life of the product is 36 months when stored at 2 to 8°C in a light-protected container. The quality and safety of the product are ensured by extensive testing.

Overview of Therapeutic Properties

Twinrix Paediatric is identical to the product marketed with the trade name Twinrix Adult, differing only in the volume dosage. It is indicated for immunization of infants, children, and adolescents up to 15 years old. Twinrix Paediatric should be administered in three intramuscular injections, with an interval of 1 month between the first and second injections and a 5-month interval between the second and third. A requirement for a booster administration has not been proven.

The combined vaccine Twinrix Paediatric was found to elicit the same immunoprotection against the hepatitis A and B viruses in children as in adults and to be as effective as the two vaccines administered separately. The side effects reported for adults were also observed in children.

Further Reading

http://www.eudra.org

http://www.gsk.com

Thoelen, S. et al., The first combined vaccine against hepatitis A and B: an overview, *Vaccine*, 17, 1657–1662, 1999.

Van Damme, P. et al., Long-term persistence of antibodies induced by vaccination and safety follow-up, with the first combined vaccine against hepatitis A and B in children and adults, *J. Med. Virol.*, 65, 6–13, 2001.

Verluma

Product Name:	Verluma (trade name) Nofetumomab (international nonproprietary name)
Description:	Verluma consists of a Fab fragment of the murine monoclonal antibody NR-LU-10, linked via a phenthiolate ligand to the radioisotope Technetium Tc 99m for radioimaging (Technetium Tc 99m Nofetumomab Merpentan). The antibody, which is produced by hybridoma technology, is specific for a glycoprotein expressed in numerous cancer and normal cells. Verluma is provided in a kit with separate components, which are necessary for the constitution of the final product before intravenous administration. The radioisotope Technetium 99m is not provided in the kit.
Approval Date:	1996 (U.S.)
Therapeutic Indications:	Verluma is indicated for diagnostic imaging in detection of extensive stage disease in patients with previously untreated, biopsy-confirmed, small-cell lung cancer.
Manufacturer:	Dr. Karl Thomae GmbH, an affiliated company of Boehringer Ingelheim International GmbH, 55216 Ingelheim, Germany, http://www.boehringer-ingelheim.com
Marketing:	DuPont Merck Pharmaceutical Company, DuPont radiopharmaceutical division, 331 Treble Cove Road, Billerica, MA 01862

Manufacturing

Verluma consists of a Fab antibody fragment obtained by enzymatic digestion of the murine IgG2b monoclonal antibody NR-LU-10. The antibody binds a 40-kDa glycoprotein expressed on the surface of numerous cancer, as well as normal, cells. The murine antibody, produced by hybridoma technology, is purified, and the Fab fragment is generated by digestion using papain. Purification includes several chromatographic steps and procedures to remove viral contaminants. The radioisotope Technetium Tc 99m is complexed with a phenthioate ligand 2,3,5,6-tetrafluorophenyl-4,5-bis-S-(1-ethoxyethyl)-thioacetoamidopentanoate and then linked to the Fab fragment to generate the product Tc 99m nofetumomab merpentan. The Verluma kit contains separate components that are necessary for the constitution of the final product: the Fab fragment nofetumomab in phosphate-buffered saline; the penthioate ligand in freeze-dried form; isopropyl alcohol; glacial acetic acid and hydrochloric acid; stannous gluconate complex containing sodium gluconate and dihydrate stannous chloride, in lyophilized form; sodium bicarbonate buffer; an empty vial; and an anion exchange column. The radioisotope Technetium Tc 99m is not provided with the kit. Isopropyl alcohol is used to dissolve the freeze-dried ligand, following which stannous ions reduce the oxidation state of Tc 99m; gluconate stabilizes the reduced Tc 99m for chelation to the ligand; sodium bicarbonate is used to adjust the pH to a basic pH for conjugation of the antibody to the complex Tc 99m-ligand; and the chromatographic column is used to eliminate the excess, unbound radioisotope and to neutralize the pH.

Verluma has a shelf life of 24 months when stored at 2 to 8°C.

Overview of Therapeutic Properties

Verluma is indicated for radiodiagnostic imaging with the radioisotope Technetium Tc 99m in detection of extensive stage disease in patients with biopsy-confirmed, untreated small-cell lung cancer. Verluma is administered as an intravenous injection and radio images are taken 14 to 17 hours after administration.

Clinical trials showed that Verluma allowed diagnosis of extensive stage disease in 85% of patients, while diagnosis of limited stage disease necessitates additional testing. Verluma failed to identify tumors in 25% of patients, in which the antigen recognized by the Fab fragment is not expressed. Verluma may lead to false-positive results if the antibody binds nontumor cells, such as excretion organs, areas of inflammation, and sites of recent surgery.

Verluma may induce the formation of human antimouse antibodies. Verluma should not be administered during pregnancy or lactation or to pediatric patients.

Further Reading

http://www.boehringer-ingelheim.com
http://www.fda.gov
Breitz, H.B. et al., Clinical experience with Tc -99m nofetumomab merpentan (Verluma) radioimmunoscintigraphy, *Clin. Nucl. Med.*, 22, 615–620, 1997.
Straka, M.R. et al., Tc-99m nofetumomab merpentan complements an equivocal bone scan for detecting skeletal metastatic disease from lung cancer, *Clin. Nucl. Med.*, 25, 54–55, 2000.

Viraferon

Product Name: Viraferon (trade name)

Interferon alfa-2b (international nonproprietary name)

Description: Viraferon is a recombinant human interferon alfa-2b. It is produced by recombinant DNA technology in *Escherichia coli.* The 19.3-kDa single chain nonglycosylated polypeptide is identical to the natural human molecule. It is provided either in a lyophilized form (containing 1, 3, 5, or 10 million IU active ingredient/vial) to be reconstituted before administration with the solvent provided or as a soluble formulation (at strengths varying from 6 to 50 million IU active/ml in vials or multidose pen cartridges). The product is administered by subcutaneous injection.

Approval Date: 2000 (E.U.)

Therapeutic Indications: Viraferon is indicated for the treatment of patients with chronic hepatitis B and chronic hepatitis C.

Manufacturer: Schering Plough (Brinny) Company, Innishannon, County Cork, Ireland, http://www.schering-plough.com

Marketing: SP Europe, 73 rue de Stalle, 1180 Bruxelles, Belgium, http://www.schering-plough.com

Manufacturing

Viraferon is a recombinant human interferon alfa-2b. The gene, derived from human leukocytes, has been cloned into a plasmid and transformed into a modified *E. coli* strain. The protein, which is identical to the natural molecule, is expressed and then extensively purified from the bacterial culture. The purification process includes chromatographic steps, crystallization, and resolubilization. The final product is presented in either a lyophilized form or as a solution in vials or in cartridges to be used with multidose devices. The lyophilized formulation contains the active substance interferon alfa-2b, glycine, dibasic sodium phosphate, monobasic sodium phosphate, and human serum albumin. The soluble formulation contains interferon alfa-2b, dibasic sodium phosphate, monobasic sodium phosphate, disodium edetate, sodium chloride, m-cresol, and polysorbate 80.

The shelf life of the product is different for the various formulations but ranges from 15 months to 3 years when stored at 2 to 8°C. The quality and safety of the product are ensured by validation tests, including SDS-PAGE, IEF, HPLC, MTT-CPE, and LAL assay.

Overview of Therapeutic Properties

Viraferon is a recombinant human interferon alfa-2b. Interferons are molecules involved in the defense against viral infections. Their antiviral action is achieved by activating molecules that inhibit viral replication, by suppressing cell proliferation, and through an immunomodulatory activity that involves inducing macrophage phagocytosis and lymphocyte cytotoxicity. Hepatitis B and C are potentially life-threatening diseases that affect a huge population worldwide. Viraferon is indicated in the treatment of chronic hepatitis B when there is evidence of viral replication, elevated liver marker enzymes, and active liver inflammation and fibrosis. It should be administered subcutaneoulsy three times a week for up to 6 months at dosages adjusted to meet the individual patient's needs. Viraferon is also indicated for the treatment of chronic hepatitis C when there is evidence of viral replication. In this case, it should be administered three times a week for up to 18 months in combination with ribavirin.

Viraferon proved its efficacy in the treatment of chronic hepatitis B and C. In the case of hepatitis C, the efficacy was increased by combined administration of ribavirin. The most commonly observed side effects were flu-like symptoms, loss of appetite, nausea, and variations in hematic and enzymatic values. Viraferon is contraindicated during pregnancy and lactation.

Further Reading

http://www.eudra.org
http://www.schering-plough.com

ViraferonPeg

Product Name:	ViraferonPeg (trade name) Peginterferon alfa-2b (international nonproprietary name)
Description:	ViraferonPeg is a recombinant modified human interferon alfa-2b. It differs from the natural molecule by the addition of a polyethylene glycol polymer strand conjugated to the molecule. It appears to be identical to the product marketed with the trade name PegIntron. ViraferonPeg is produced by recombinant DNA technology in *Escherichia coli* cells and is presented as a powder (at strengths of 50 to 150 μg/dose) to be resuspended before subcutaneous injection. It is also provided in vials and cartridges to be used with multidose devices.
Approval Date:	2000 (E.U.)
Therapeutic Indications:	ViraferonPeg is indicated for the treatment of adult patients with chronic hepatitis C and is usually administered in combination with Ribavirin.
Manufacturer:	Schering Plough (Brinny) Company, Innishannon, County Cork, Ireland, http://www.schering-plough.com (manufacturer is responsible for batch release in the European Economic area)
Marketing:	SP Europe, 73 rue de Stalle, 1180 Bruxelles, Belgium, http://www.schering-plough.com

Manufacturing

ViraferonPeg is a recombinant, PEGylated human interferon alfa-2b. The interferon gene, derived from human leukocytes, has been cloned into a plasmid and transformed into a modified *E. coli* strain. The protein, which is identical to the natural molecule, is expressed and purified from the bacterial culture (see also the monographs for Viraferon and PegIntron). As part of the downstream processing, the interferon is incubated with activated methoxypolyethylene glycol (mPEG), resulting in formation of a covalent linkage via protein amino groups. Subsequent characterization by mass spectroscopy, SDS-PAGE, and size exclusion chromatography indicate that the product consists predominantly of a monopegylated species, with minor quantities of both unpegylated and dipegylated moieties present.

The lyophilized formulation contains PEGylated interferon alfa-2b (active), as well as dibasic sodium phosphate, monobasic sodium phosphate, sucrose, and polysorbate 80 as excipients.

Overview of Therapeutic Properties

ViraferonPeg is indicated in the treatment of adult patients with histologically proven chronic hepatitis C who have elevated transaminase levels without liver decompensation and who are positive for serum hepatitis C virus RNA/antibodies. PEGylation of the interferon results in extending its plasma half-life by decreasing the rate of systemic clearance without affecting its antiviral properties. The product is administered only once a week by subcutaneous injection and usually in combination with ribavirin. Treatment usually lasts for 6 months, although it may be extended if necessary.

The most common side effects included flu-like symptoms, nausea, anxiety and depression, and chest pain. ViraferonPeg is contraindicated during pregnancy and lactation. (See also the monographs for Viraferon and PegIntron.)

Further Reading

http://www.eudra.org

http://www.schering-plough.com

Haugh, P. et al., Viraferon Peg plus ribavirin treatment in patients with compensated HCV cirrosis, *Gut*, 52(5), 68–68, 2003.

Vitravene (withdrawn from E.U. market)

Product Name: Vitravene (trade name)

Fomivirsen (international nonproprietary name)

Description: The active ingredient in Vitravene is a 21-base phosphorothioate oligonucleotide of the indicated nucleotide base sequence 5′ - GCG TTT GCT CTT CTT CTT GCG - 3′. It is a white to off-white hygroscopic, amorphous powder with a molecular formula of $C_{204}H_{243}N_{63}O_{114}P_{20}S_{20}Na_{20}$ and a molecular weight of 7122 Da. It is the only antisense-based product approved in any world region thus far. The final product is supplied in a preservative-free, single-use vial containing 0.25 ml. In addition to the active ingredient (at a concentration of 26.4 mg/ml), the product contains sodium carbonate, sodium bicarbonate, and sodium chloride formulated in water for injection. The final product pH is adjusted to a pH of 8.7 with NaOH/HCl.

Approval Date: 1998 (U.S.); 1999 (E.U.)

Withdrawal Date: Vitravene was withdrawn from the European market in 2002 for commercial reasons. The product is still marketed in Switzerland, from where it may be supplied to European member states on a named patient basis. The product remains available in the U.S. market.

Therapeutic Indications: Vitravene is indicated for the local treatment of cytomegalovirus (CMV) retinitis in patients with acquired immunodeficiency syndrome (AIDS) who are intolerant, have a contraindication to other treatments for CMV retinitis, or were insufficiently responsive to previous treatments for CMV retinitis.

Manufacturer: Isis Pharmaceuticals Inc., 2292 Faraday Avenue, Carlsbad, CA 92008-7208, http://www.isispharm.com

Marketing: Novartis Ophthalmics (formerly CIBA Vision), One Health Plaza, East Hanover, NJ 07936, http://www.novartisophthalmics.com

Manufacturing

The active ingredient in Vitravene is manufactured by direct chemical synthesis. The active is then formulated with the excipients, sterilized, and filled into glass vials.

Overview of Therapeutic Properties

Vitravene is unique in that it appears to be the only nucleic acid-based biopharmaceutical approved for general medical use thus far. Antisense technology is based upon the generation of short single-stranded stretches of nucleic acids of predefined nucleotide base sequence. The nucleotides are capable of binding either to specific genes or, more commonly, to mRNA derived from specific genes. Binding is usually promoted via the Watson–Crick-based nucleotide base complementarity. Binding, therefore, prevents gene expression by inhibiting either gene transcription or (more commonly) gene translation. Down-regulation/inhibition of gene expression could be important in treating medical conditions caused or exacerbated by inappropriate gene expression or overexpression. Most nucleic acids used for antisense purposes are manufactured from chemically modified nucleotides in order to make them more resistant to degradation by nucleases. Phosphorothioate-based oligonucleotides are most commonly used.

The phosphorothioate oligonucleotide present in Vitravene displays a base sequence complementary to a sequence of mRNA transcripts of the major immediate early region-2 (IE-2) of human cytomegalovirus (CMV). This region of mRNA encodes several proteins responsible for regulation of viral gene expression that are essential for production of infectious CMV. Binding of the active ingredient to target mRNA inhibits IE-2 protein synthesis via an antisense mechanism, thereby inhibiting viral replication.

Treatment with Vitravene entails direct intravitreal injection (330 µg in 0.05 ml) initially every second week and subsequently once every 4 weeks. The most frequently observed side effect is ocular inflammation, which typically occurs in one of every four patients.

Further Reading

Levin, A.A., A review of issues in the pharmacokinetics and toxicology of phosphorothiolate antisense oligonucleotides, *BBA – Gene Str. Expr.*, 1489(1), 69–84, 1999.

Lysik, M.A. and Wu-Pong, S., Innovations in oligonucleotide drug delivery, *J. Pharm. Sci.*, 92(8), 1559–1573, 2003.

Meenken, C. et al., Intravitreal treatment of cytomegalovirus (CMV) retinitis with formivirsen (vitravene) in patients with AIDS, *AIDS*, 14, 389, 2000.
Reese, C.B. and Yan, H.B., Solution phase synthesis of ISIS 2922 (vitravene) by the modified H-phosphonate approach, *J. Chem. Soc. Perk. T.*, 1(23), 2619–2633, 2002.

Xigris

Product Name: Xigris (trade name)

Drotrecogin alfa (activated) (international nonproprietary name)

Description: Xigris consists of a recombinant human activated protein C, an antithrombotic molecule very similar to the natural human molecule that regulates coagulation. Xigris is produced in a modified human cell line using recombinant DNA technology. It is a 55-kDa glycosylated serine protease consisting of two polypeptide chains (a light and a heavy chain) linked by a disulphide bond. Nine glutamic acid residues present on the light chain are γ-carboxylated and one aspartic acid residue is β-hydroxylated. Xigris is supplied as a lyophilized form (5 mg/vial or 20 mg/vial) to be resuspended before intravenous infusion.

Approval Date: 2001 (U.S.); 2002 (E.U.)

Therapeutic Indications: Xigris is indicated for the treatment of adult patients with severe sepsis associated with multiple organ failure who are at high risk of death.

Manufacturer: Lilly Pharma Fertigung und Distribution GmbH and Co. KG, Teichweg 3, 35396 Giessen, Germany, http://www.lilly.com (manufacturer is responsible for import and batch release in the European Economic Area)
Eli Lilly and Company, Lilly Corporate Center, Indianapolis, IN 46285, http://www.lilly.com (U.S.)

Marketing: Eli Lilly Nederland B.V., Grootslag 1-5, 3991 RA, Houten, the Netherlands, http://www.lilly.com (E.U.)
Eli Lilly and Company, Lilly Corporate Center, Indianapolis, IN 46285, http://www.lilly.com (U.S.)

Manufacturing

Drotrecogin alfa (activated) is a recombinant human protein C produced in a modified human cell line, HEK293, using recombinant DNA technology. The human protein C gene has been obtained from cDNA from the liver of a healthy donor. The recombinant protein C, or zymogen, the precursor of the active form, is purified from the culture using chromatographic procedures. The purified product is converted into the active form using thrombin of bovine origin. The molecule is further purified and subjected to procedures for inactivation and removal of viral contaminants. The recombinant-activated protein C differs from the natural molecule only in its glycosylation pattern. Xigris consists of the active substance drotrecogin alfa (activated), sucrose (as a bulking agent and stabilizer for the solid state), sodium chloride (as a bulking agent, stabilizer for the solution state), sodium citrate (as a buffering agent), citric acid, hydrochloric acid, and sodium hydroxide.

The shelf life of the product is 24 months when stored at a temperature of 2 to 8°C, protected from light. Control tests are carried out to ensure the quality and safety of the product, including extensive procedures to test for the presence of viral contamination.

Overview of Therapeutic Properties

The recombinant human activated protein C is very similar to the natural human molecule, an antithrombotic agent involved in the regulation of coagulation and inflammation. Severe infection may cause formation of blood clots, leading to severe sepsis, when the blood supply to vital organs is blocked. This can lead to organ failure, with life-threatening consequences. Xigris is administered in order to eliminate blood clots and reduce the inflammation caused by the infection. It achieves this largely by inactivating the blood coagulation factors V_a and VIII_a and by inhibiting plasminogen activator inhibitor-1 (PAI-1), thereby enhancing the fibrinolytic response. Xigris should be administered as an intravenous infusion over a period of 96 hours to adult patients at high risk of death. Clinical studies showed that patients treated with Xigris experience an improved 28-day survival compared to patients receiving placebos. The reduction in mortality was observed only in patients with a great degree of disease severity who were suffering from severe acute organ dysfunctions.

The most common side effect reported after administration of Xigris was bleeding, and headaches were also reported. Severe bleeding was observed in patients at high risk of bleeding, to whom Xigris should not be administered. Extreme care should be taken if patients are administered drugs that affect hemostasis. Xigris should not be administered to patients with severe liver disease, with low platelet count, to children younger than 18 years old, and during pregnancy and lactation.

Further Reading

http://www.eudra.org
http://www.fda.gov
http://www.lilly.com
http://www.xigris.com
Bernard, G.R. et al., Efficacy and safety of recombinant human activated protein C for severe sepsis, *N. Engl. J. Med.*, 344, 699–709, 2001.
Lyseng-Williamson, K.A. and Perry, C.M., Drotrecogin alfa (activated), *Drugs*, 62, 617–630, 2002.
Macias, W.L. et al., Pharmacokinetic-pharmacodynamic analysis of drotrecogin alfa (activated) in patients with severe sepsis, *Clin. Pharmacol. Ther.*, 72, 391–402, 2002.

Xolair

Product Name: Xolair (trade name)

Omalizumab (common name)

Description: The active ingredient of Xolair is a humanized IgG1 κ monoclonal antibody, which selectively binds to immunoglobulin E (IgE). The 149-kDa antibody is produced in a mammalian cell line and is presented as a sterile, preservative-free lyophilized powder destined for subcutaneous administration subsequent to reconstitution in a water-for-injection–based solvent. Reconstitution yields an active ingredient concentration of 125 mg/ml.

Approval Date: 2003 (U.S.)

Therapeutic Indications: Xolair is indicated for the treatment of adults and adolescents (12 years or older) with moderate to severe persistent asthma, who have a positive skin test or *in vitro* reactivity to a perennial aeroallergen, and whose symptoms are inadequately controlled with inhaled corticosteroids. Xolair has been shown to decrease the incidence of asthma exacerbations in these patients.

Manufacturer: Genentech Inc., 1 DNA Way, South San Francisco, CA 94080-4990, http://www.gene.com

Marketing: Genentech Inc., 1 DNA Way, South San Francisco, CA 94080-4990, http://www.gene.com; and Novartis Pharmaceuticals Corporation, One Health Plaza, East Hanover, NJ 07936-1080, http://www.novartis.com

Manufacturing

Omalizumab is a humanized monoclonal antibody produced by recombinant DNA technology in an engineered Chinese hamster ovary (CHO) cell line. The producer cells are grown in a suspension culture, in a nutrient medium containing gentamicin, although no gentamicin is present in the final product. Downstream processing is initiated by cellular harvest and recovery of the extracellular antibody. Multistep chromatographic purification ensues, and the production protocol includes a viral inactivation and removal step. Sucrose, L-histidine, and polysorbate 20 are added as excipients. The final product is sterilized by filtration, filled into single-use vials, and lyophilized.

Overview of Therapeutic Properties

Xolair brings about its therapeutic effect by binding to IgE, a major mediator of many allergic reactions. Xolair-bound IgE fails to bind to its high affinity IgE receptor (FcεR1) on the surface of mast cells and basophils; therefore the IgE-triggered release of mediators of the allergic response form those cells is limited. Once bound to Omalizumab, IgE is proposed to be cleared from the body via macrophage endocytotic clearance. Treatment also reduces the number of FcεR1 receptors on the surface of basophils in atopic patients.

Xolair is typically administered subcutaneously to its target population every 2 or 4 weeks in doses equivalent to between 150 and 375 mg active substance. Exact dosage levels and regimens are determined by starting total serum IgE level (IU/ml) and body weight. The lyophilized product takes 15 to 20 minutes to fully dissolve in the solvent provided (water for injection). The resultant reconstituted product is somewhat viscous, rendering necessary careful withdrawal from the vial into the syringe as well as slightly prolonged duration of administration. Clinical trials generally showed that the number of exacerbations (worsening of asthma, necessitating treatment with systemic corticosteroids, etc.) per patient decreased in patients treated with Xolair. Xolair has not been shown to alleviate asthma exacerbations acutely and should not be used for the treatment of acute bronchospasm or status asthmaticus. Systemic or inhaled corticosteroids should not be abruptly discontinued upon initiation of Xolair therapy.

Administration of Xolair appeared to increase the incidence of malignant neoplasms (0.5% in Xolair-treated patient, as compared to 0.2% in control patients). Anaphylactic reactions were also noted in fewer than 0.1% of patients. Among the more common side effects were viral and upper respiratory tract infections, headaches, and pharyngitis. Safety and effectiveness in pediatric patients younger than 12 years of age have not been established. As IgG is excreted naturally in milk, it is likely that Xolair will be present in the milk of nursing mothers, and the product should be administered to such women with caution. The shelf life of the product is 18 months when stored at 2 to 8°C.

Further Reading

Anon., Omalizumab (Xolair): an anti -IgE antibody for asthma, *Med. Lett. Drugs Ther.*, 45(1163), 67–68, 2003.

Babu, K.S. et al., Omalizumab, a novel anti -IgE therapy in allergic disorders, *Expert Opin. Biol. Th.*, 1(6), 1049–1058, 2001.

Cullenn-Young, M. et al., Omalizumab. Treatment of allergic rhinitis, treatment of asthma, *Drug Future*, 27(6) 537–545, 2002.

Hochhaus, G. et al., Pharmacodynamics of Omalizumab: implications for optimized dosing strategies and clinical efficacy in the treatment of allergic asthma, *Curr. Med. Res. Opin.*, 19(6), 491–498, 2003.

Owen, C.E., Anti-immunoglobulin E therapy for asthma, *Pulm. Pharmacol. Ther.*, 15(5), 417–424, 2002.

Zenapax

Product Name:	Zenapax (trade name) Daclizumab (international nonproprietary name)
Description:	Zenapax is a humanized monoclonal antibody with specific binding activity for the α-chain (also known as CD25 or Tac) of the interleukin-2 (IL-2) receptor expressed on the surface of activated lymphocytes. It is produced using recombinant DNA technology in NSO myeloma cells and supplied as a concentrated solution (5 mg/ml) for intravenous infusion.
Approval Date:	1997 (U.S.); 1999 (E.U.)
Therapeutic Indications:	Zenapax is used as an immunosuppressant for the prophylaxis of acute organ rejection in patients receiving *de novo* allogeneic renal transplants. It is used in combination with other immunosuppressive drugs, including cyclosporine and corticosteroids.
Manufacturer:	Hoffmann-La Roche AG, Postfach 1270, 79630 Grenzach-Wyhlen, Germany (manufacturer is responsible for import and batch release in the European Economic Area) Hoffmann-La Roche Inc., 340 Kingsland Street, Nutley, NJ 07110, http://www.rocheusa.com (U.S.)
Marketing:	Roche Registration Limited, 40 Broadwater Road, Welwyn Garden City, Hertfordshire, AL7 3AY, U.K. (E.U.) Hoffmann-La Roche Inc., 340 Kingsland Street, Nutley, NJ 07110, http://www.rocheusa.com (U.S.)

Manufacturing

The hybridoma cells producing the monoclonal antibody were obtained by fusion of the NS-1 mouse myeloma cell line to spleen cells from a mouse immunized against human T cells. The recombinant humanized antibody was constructed by combining the complementarity-determining regions (CDRs) of the mouse monoclonal antibody with human framework (from the Eu myeloma antibody) and constant domains (from human IgG1). Computer modeling of the mouse antibody revealed that amino acid residues outside the CDRs were likely influencing binding, and these residues were maintained in the humanized antibody.

The murine GS-NSO myeloma cell line is used to produce the recombinant antibody in a bioreactor. The purification process consists of a combination of chromatographic steps leading to the purified product. The final product contains daclizumab (active), as well as polysorbate 80, sodium phosphate monobasic and dibasic, sodium chloride, and hydrochloric acid or sodium hydroxide as excipients. The shelf life of the product is 24 months when stored at 2 to 8°C.

Routine evaluative tests were carried out on the final product for physical specifications, identity and purity (peptide map, isoelectric focusing, SE-HPLC, SDS-PAGE, Western blotting, size exclusion chromatography, N-terminal sequence analysis, IL-2 binding assay), for the presence of endotoxins, microbiological and viral safety, and functionality of the protein (bioassay).

Overview of Therapeutic Properties

The humanized recombinant antibody Zenapax binds specifically the α-chain of the IL-2 receptor on the surface of activated lymphocytes. It acts as an antagonist of the receptor, blocking the binding of IL-2 and preventing the stimulation of lymphocytes involved in organ rejection.

Five doses of the product are administered to the patient by 20-minute infusions. Administration is every 2 weeks, with the first dose at the time of the transplant. Zenapax, with a half-life similar to an IgG (about 20 days), saturates the receptor for up to 4 months after the last dose. Clinical studies on the use of Zenapax as part of a immunosuppressive regimen showed that it reduced the incidence of biopsy-proven rejection within 6 months of transplantation from 47 to 28% in a double-therapy study (Zenapax with cyclosporine and corticosteroids) and from 35 to 22% in a triple-therapy study (Zenapax with cyclosporine, corticosteroids, and azathioprine). Patient survival at 1-year post-transplant increased in the double-therapy study when Zenapax was administered. Zenapax did not increase the incidence of side effects that are usually observed with the standard immunosuppressive regimen, the most common being gastrointestinal disorders and infections. The humanized form of the antibody (90% human and 10% murine) has the advantage over the murine antibody of reduced immunogenicity while maintaining the binding activity. Significant hypersensitivity reactions were only rarely observed.

Because of the potential risks to the fetus, Zenapax is contraindicated during pregnancy and should be avoided during lactation. Very few data are available on the use of Zenapax for pediatric patients. Studies of the use of Zenapax for treating autoimmune diseases are currently underway.

Further Reading

http://www.eudra.org
http://www.fda.gov
http://www.roche.com
Bell, J. and Colaneri, J., Zenapax: transplant's first humanized monoclonal antibody, *ANNA J.*, 25, 429–430, 1998.
Vincenti, F. et al., Interleukin-2-receptor blockade with daclizumab to prevent acute rejection in renal transplantation. Daclizumab Triple Therapy Study Group, *N. Engl. J. Med.*, 338, 161–165, 1998.
Wiseman, L.R. et al., Daclizumab: a review of its use in the prevention of acute rejection in renal transplant recipients, *Drugs*, 58, 1029–1042, 1999.

Zevalin

Product Name: Zevalin (trade name)

Ibritumomab tiuxetan (international nonproprietary name)

Description: Ibritumomab is a murine derived IgG1 kappa monoclonal antibody specific for the human CD20 antigen, which is expressed on the surface of normal and malignant B cells. Ibritumomab is used coupled to the radioisotopes Indium-111 (In-111) and Yttrium-90 (Y-90), via the linker-chelator tiurexan and in combination with the recombinant chimeric monoclonal antibody Rituximab. Ibritumomab is produced in Chinese hamster ovary (CHO) cells using recombinant DNA technology. Zevalin is presented in separate kits for the preparation of intravenous administration of In-111 ibritumomab tiuxetan or Y-90 ibritumomab tiuxetan.

Approval Date: 2002 (U.S.)

Therapeutic Indications: Zevalin is indicated for the treatment of patients with relapsed or refractory low-grade, follicular, or transformed B-cell non-Hodgkin's lymphoma, including patients with Rituximab refractory follicular non-Hodgkin's lymphoma.

Manufacturer: IDEC Pharmaceuticals Corporation, 3030 Callan Road, San Diego, CA 92121, http://www.idecpharm.com

Marketing: IDEC Pharmaceuticals Corporation, 3030 Callan Road, San Diego, CA 92121, http://www.idecpharm.com

Manufacturing

The murine IgG1 kappa monoclonal antibody ibritumomab is produced in CHO cells using recombinant DNA technology. The antibody is specific for the human CD20 antigen that is expressed on the surface of normal and malignant B lymphocytes. After a multistep chromatographic purification, the antibody is covalently coupled to the linker-chelator tiurexan, [N-[2-bis(carboxymethyl)amino]-3-(p-isothiocyanatophenyl)-propyl]-[N-[2-bis(carboxymethyl)amino]-2-(methyl)-ethyl]glycine, which exhibits a chelation site for In-111 or Y-90. Zevalin is provided in two separate kits for use with In-111 or Y-90, each containing ibritumomab tiuxetan in a sodium chloride solution; sodium acetate trihydrate; a solution of human albumin, sodium chloride, dibasic heptahydrate sodium phosphate, pentetic acid, monobasic potassium phosphate, potassium chloride, sodium chloride, and hydrochloridric acid; and a receptacle for the preparation of the product before administration. The radioisotopes are not provided with the kits.

The shelf life of the product is 24 months when stored at 2 to 8°C.

Overview of Therapeutic Properties

Low-grade or follicular non-Hodgkin's lymphoma is an incurable form of cancer. Patients may exhibit disease remission for years, but ultimately relapses become more frequent and responses to treatment become less effective over time. Zevalin is indicated in the treatment of patients with relapsed or refractory low-grade, follicular, or transformed B-cell non-Hodgkin's lymphoma, including patients with Rituximab refractory follicular non-Hodgkin's lymphoma. Zevalin is the first treatment that combines the use of monoclonal antibodies and radioactive compounds. The antibody moiety targets the human CD20 antigen, which is expressed on the surface of normal and malignant B cells and is coupled to the radioisotope Y-90 which induces cell death. Zevalin is administered in combination with Rituximab. The treatment includes the use of the antibody coupled to the radioisotope In-111 for radioimaging. Intravenous administration of In-111 Zevalin, within 4 hours of a 2-hour infusion of Rituximab, is followed by gamma imaging to assess the biodistribution of the radioisotope. An initial image is taken 2 to 24 hours after injection of In-111 Zevalin, a second is taken between 48 and 72 hours after the injection, and a third is taken 90 to 120 hours after the injection if required. The treatment should be discontinued in the case of an altered biodistribution. Otherwise, an infusion of Rituximab is repeated 7 to 9 days after the injection of In-111 Zevalin, followed within 4 hours by an intravenous injection of Y-90 Zevalin.

Clinical studies showed an overall response of 75 to 80% following Zevalin treatment in patients who no longer responded to Rituximab or chemotherapy. Zevalin treatment showed a better overall response and a response duration 2 months longer than that observed with Rituximab. Survival rates are not yet available.

Zevalin therapy was found to be more toxic than Rituximab therapy, with severe reduction in white blood cell counts lasting 12 weeks from the end of treatment and returning to normal 9 months after treatment. A severe reduction in the blood platelets was also reported. Fatal severe infusion-related reactions, infection, and hemorrhagic episodes were reported. Malignancies developed in 2% of patients. Zevalin is contraindicated during pregnancy and lactation.

Further Reading

http://www.fda.gov

Ihttp://www.idecpharm.com

http://www.zevalin.com

Alcindor, T. and Witzig, T.E., Radioimmunotherapy with yttrium-90 ibritumomab tiuxetan for patients with relapsed CD20+ B-cell non-Hodgkin's lymphoma, *Curr. Treat. Options Oncol.*, 3, 275–282, 2002.

Juweid, M.E., Radioimmunotherapy of B-cell non-Hodgkin's lymphoma: from clinical trials to clinical practice, *J. Nucl. Med.*, 43, 1507–1529, 2002.

Wiseman, G.A. et al., Ibritumomab tiuxetan radioimmunotherapy for patients with relapsed or refractory non-Hodgkin's lymphoma and mild thrombocytopenia: a phase II multicenter trial, *Blood*, 99, 4336–4342, 2002.

Witzig, T.E. et al., Treatment with ibritumomab tiuxetan radioimmunotherapy in patients with rituximab-refractory follicular non-Hodgkin's lymphoma, *J. Clin. Oncol.*, 20, 3262–3269, 2002.

Witzig, T.E. et al., Randomized controlled trial of yttrium-90-labeled ibritumomab tiuxetan radioimmunotherapy versus rituximab immunotherapy for patients with relapsed or refractory low-grade, follicular, or transformed B-cell non-Hodgkin's lymphoma, *J. Clin. Oncol.*, 20, 2453–2463, 2002.

Appendix

Monographs are grouped by biological activity.

Antibody-Based Products

Bexxar
CEA-Scan
Herceptin
HumaSPECT
Humira
Indimacis 125
LeukoScan
MabCampath/Campath-1H
MabThera/Rituxan
Mylotarg
Myoscint
OncoScint CR/OV
Orthoclone OKT-3
ProstaScint
Remicade
ReoPro
Simulect
Synagis
Tecnemab K1
Verluma
Xolair
Zenapax
Zevalin

Anticoagulants

Refludan
Revasc
Xigris

Antisense-Based Products

Vitravene

Blood Factors

BeneFIX
Bioclate
Helixate NexGen/Helixate FS/Kogenate FS
Novoseven
Recombinate
ReFacto

Bone Morphogenetic Proteins

InductOs
InFUSE Bone Graft/LT-CAGE Lumbar Tapered Fusion Device
Osigraft
OP-1 Implant

Colony Stimulating Factors

Leukine
Neulasta
Neupogen

Erythropoietin

Epogen/Procrit
NeoRecormon
Nespo/Aranesp

Fusion Products

Amevive
Enbrel
Ontak

Growth Factors

Regranex

Hormones

Bio-Tropin (Tev-Tropin or Zomacton)
Forcaltonin
Forsteo/Forteo
Genotropin
GlucaGen
Glucagon
Gonal-F
Humalog
Humatrope
Humulin
Insuman
Lantus
Liprolog
Luveris
Natrecor
Norditropin
Novolin/Actrapid/Insulatard/Mixtard/Monotard/Ultratard/Velosulin
NovoRapid/Novolog and NovoMix 30/NovoLog Mix 70/30
Nutropin Depot
NutropinAQ
Optisulin
Ovitrelle/Ovidrel
Protropin
Puregon/Follistim
Saizen
Serostim
Somavert
Thyrogen

Interferons

Actimmune
Avonex
Betaferon/Betaseron
Infergen
Intron A
Pegasys
PegIntron
Rebetron
Rebif
Roferon A
Viraferon
ViraferonPeg

Interleukins, Tumor Necrosis Factors, and Related Products

Beromun
Kineret
Neumega
Proleukin

Therapeutic Enzymes

Aldurazyme
Cerezyme
Fabrazyme
Fasturtec/Elitek
Pulmozyme
Replagal

Thrombolytic Agents

Activase
Ecokinase
Metalyse/TNKase
Rapilysin/Retavase

Vaccines

Ambirix
Comvax
Engerix-B
HBVAXPRO
Hepacare
Hexavac
INFANRIX HepB
Infanrix Hexa
Infanrix Penta/Pediarix
LYMErix
Primavax
PROCOMVAX/COMVAX
Recombivax
Triacelluvax
Tritanrix HepB
Twinrix Adult
Twinrix Paediatric